SpringerBriefs in Applied Sciences and Technology

PoliMI SpringerBriefs

Springer, in cooperation with Politecnico di Milano, publishes the PoliMI Springer-Briefs, concise summaries of cutting-edge research and practical applications across a wide spectrum of fields. Featuring compact volumes of 50 to 125 (150 as a maximum) pages, the series covers a range of contents from professional to academic in the following research areas carried out at Politecnico:

- Aerospace Engineering
- Bioengineering
- Electrical Engineering
- Energy and Nuclear Science and Technology
- Environmental and Infrastructure Engineering
- Industrial Chemistry and Chemical Engineering
- Information Technology
- Management, Economics and Industrial Engineering
- Materials Engineering
- Mathematical Models and Methods in Engineering
- Mechanical Engineering
- Structural Seismic and Geotechnical Engineering
- Built Environment and Construction Engineering
- Physics
- Design and Technologies
- Urban Planning, Design, and Policy

http://www.polimi.it

Bruno Daniotti · Sonia Lupica Spagnolo ·
Alberto Pavan · Cecilia Maria Bolognesi
Editors

Innovative Tools
and Methods Using BIM
for an Efficient Renovation
in Buildings

POLITECNICO
MILANO 1863

Editors
Bruno Daniotti
Department ABC—Architecture, Built
Environment and Construction Engineering
Politecnico di Milano
Milano, Italy

Sonia Lupica Spagnolo
Department ABC—Architecture, Built
Environment and Construction Engineering
Politecnico di Milano
Milano, Italy

Alberto Pavan
Department ABC—Architecture, Built
Environment and Construction Engineering
Politecnico di Milano
Milano, Italy

Cecilia Maria Bolognesi
Department ABC—Architecture, Built
Environment and Construction Engineering
Politecnico di Milano
Milano, Italy

ISSN 2191-530X   ISSN 2191-5318 (electronic)
SpringerBriefs in Applied Sciences and Technology
ISSN 2282-2577   ISSN 2282-2585 (electronic)
PoliMI SpringerBriefs
ISBN 978-3-031-04669-8   ISBN 978-3-031-04670-4 (eBook)
https://doi.org/10.1007/978-3-031-04670-4

This Springer imprint is published by the registered company Springer Nature Switzerland AG
The registered company address is: Gewerbestrasse 11, 6330 Cham, Switzerland

Preface

Since the last decades of past century, a change of paradigm entered the construction sector: the introduction of Building Information Modelling (BIM) has given the sector new opportunities to enrich the whole process with new potentialities for the relevant stakeholders. In particular the "Information" managed at the level of projects' objects allowed to apply a set of methods developed in the past century which still were pending as theoretical: for example, the systemic and performance-based approaches has found in the ICT developments the perfect ground to pass from theory to practice. Within this framework this EU project BIM4EEB, represents a step focused on the development of a BIM based toolkit for building renovation, which is fundamental for Europe in a period during which there is the need to recover existing buildings, rather than constructing new ones, considering sustainability development goals.

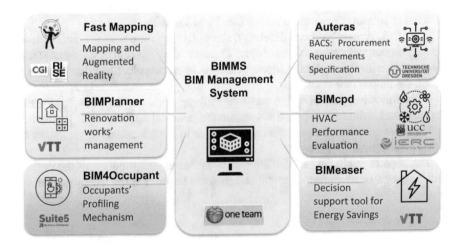

Fig. 1 BIM4EEB BIM based toolkit

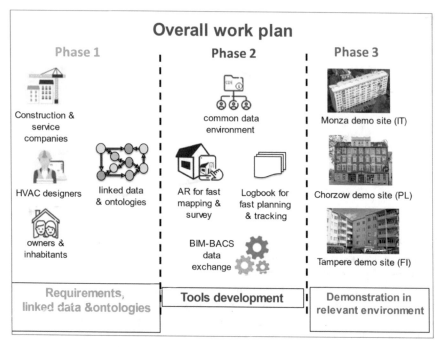

Fig. 2 BIM4EEB overall work plan

Specifically, BIM4EEB has been developed with the intention to make the flow of information efficient, to decrease intervention working time and costs, and to improve building performances, with a specific focus on quality and comfort for inhabitants. In this book we report the chief BIM4EEB results represented by the BIM bases buildings renovation toolkit (see Figs. 1 and 2).

BIM4EEB project has been developed since 2019, starting with the definition of the information requirements (see chapter 2) and the development of specific ontologies to deal with linked data to support efficient interoperable, open data exchange within the toolkit (see chapter 3).

In the following chapter 4, 5, 6 and 7 the different tools developed by BIM4EEB are presented, while in chapter 8 it's shown the practical application of the toolkit to 3 demo sites (Italy, Poland, Finland) in order to have a complete validation of BIM4EEB results.

The hope in publishing BIM4EEB results is that the contribution of the project will be useful in supporting relevant stakeholders improving the whole renovation process.

Milano, Italy   Bruno Daniotti

Contents

Chapter 1
Information Requirements for an Efficient Renovation Process

Sonia Lupica Spagnolo, Martina Signorini, Teemu Vesanen, Alberto Pavan, and Spiros Kousouris

Abstract When a renovation process takes place, different stakeholders are responsible of several activities and their interaction occurs at different stages of the building process. Therefore, a deep analysis of possible activities for different stakeholders in each different stage of the life cycle helps outlining how to optimize their interaction, thanks to the use of BIM-based tools that can smooth collaboration and data gathering. As it is commonly agreed that information losses, data lacks or redundancies are one of the main causes of time delay and cost increase, a flowchart representing the building process in case of renovation has been developed and then used to design a BIM management system (BIMMS) to allow every stakeholder along the life cycle of a building (and built asset) finding required information and share existing or new datasets in a straightforward and conflict-free manner. A particular attention has been paid in individualising differences between the public and private sectors, to be successfully applied to the renovation process in both the sectors.

Keywords Renovation processes · Stakeholders · Information requirements

1.1 The Information Workflow in a Renovation Process

When a renovation process takes place, different stakeholders are responsible of several activities. For mapping a renovation process, first of all, a list of involved stakeholders has been produced considering also the literature review (CCC 2006;

S. Lupica Spagnolo (✉) · M. Signorini · A. Pavan
Department ABC—Architecture, Built Environment and Construction Engineering, Politecnico di Milano, Milano, Italy
e-mail: sonia.lupica@polimi.it

T. Vesanen
VTT Technical Research Centre of Finland Ltd., Espoo, Finland

S. Kousouris
SUITE5 Data Intelligence Solutions Limited, Limassol, Cyprus

© The Author(s) 2022
B. Daniotti et al. (eds.), *Innovative Tools and Methods Using BIM for an Efficient Renovation in Buildings*, SpringerBriefs in Applied Sciences and Technology, https://doi.org/10.1007/978-3-031-04670-4_1

GBPN 2013; RIBA 2013). Furthermore, stages of building processes have been defined, according to (EN 16310:2013), (ISO 6707-1:2020) and (ISO 6707-2:2017).

The process of renovation needs a methodology that takes into consideration several parameters such as the cost of the renovation measures, the targeted performance (Key Performance Indicators like indoor thermal comfort levels, CO_2-emissions, energy, safety), national building regulations, building specific information and the end user needs. Also, the process must be adjusted to allow some iterations, because the planning and the implementation of the renovation measures and the conditions surveys during the process might bring up some hidden deterioration of the building components having an effect either positively or negatively on the targeted performance of the renovation actions.

By listing all the possible actors involved along a renovation process, information workflow can be defined to be considered when developing the BIM Management system to assure a proper definition of ontologies and interoperable exchange formats.

The output of this deed analysis is a flowchart representing actors and actions where:

- actions are represented by rectangular shapes,
- dotted lines are used for actions that are used for actions that are not common to all analysed countries,
- lilac rectangles represent the actions made only in public works,
- decision gates are represented by diamond shapes.

These activities are the starting point for mapping outputs of such activities (that could be mapped in a traditional process and in BIM-based processes) and for estimating times spent for carrying activities and achieving defined outputs. The synoptic flowchart will be used to outline how ICT and BIM can support the rationalisation of information flows (BIM4EEB D2.1, 2019) (Fig. 1.1).

1.2 Definition of Designers' Needs and Requirements for BIM-Assisted Decision Support for Performance-Based Refurbishment

The previous paragraph presents the renovation process described in the project from the general point of view. When the BIM4EEB project proceeded further, that process was updated from the designers' point of view related to the new activities and roles found. A general view of the designer requirements and the concept of performance-based decision making with a specific use case of the BIM-assisted scenario simulator is shown.

The model and energy calculation gradually refine in the energy performance design from a simple monthly calculation with basic information into a dynamic simulation with an accurate building information model (BIM) and detailed information about technical building systems. Figure 1.2 illustrates the scope of the work.

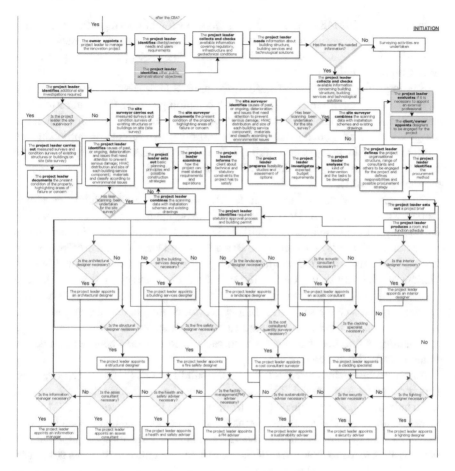

Fig. 1.1 Extract of the developed flowchart for a renovation process

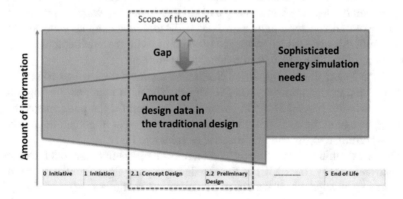

Fig. 1.2 The information gap between the early-stage information need

In this task in the BIM4EEB project the requirements for the data regarding successful performance-based design were studied. This chapter is based on the project deliverable (BIM4EEB D2.2, 2019), which is available in the project website https://www. bim4eeb-project.eu/.

Building Energy simulation (BES) tools are used for analysing energy need and expected indoor environmental quality in the buildings under different retrofitting plans. It essential to analyse not only the energy, but also the value for which the energy is used. We are not interested to just conserving energy, but also optimising the comfort and productivity in the buildings. Typically, BES can calculate performance indicators such as CO_2 emissions, LCC, LCA, occupant's satisfaction and indoor environment quality.

On a general level, the input data to BES are the weather, building envelope, geometry and orientation, HVAC system and the occupant's behaviour. The heating and cooling set-points and the air ventilation rates are the most influential parameters on the building's energy consumption. Besides, in cold weathers, main effects on the heating load come from the building envelope parameters of the window U-value, window g-value and wall conductivity. In hot weathers, the cooling load is mainly affected by the solar-heat gains, the internal heat gains from the occupant's behaviour, and the heaviness of the structures. In addition to its influence on the output results of BES, occupant's behaviour related to people existence and use of appliances and lighting is one major factor that can cause discrepancies between simulation prediction and the real energy use.

A first use case in the project is to model the building as it is before the renovations. This is called an As-is model. The accurate acquisition of data for the As-is models of the buildings that are subjected to retrofitting is the first challenge for BIM-based energy simulations. Old buildings normally lack complete and updated documentation of all data that are required by energy simulation. Appropriate equipment and methods are needed to monitor, collect, and manage the data and the tools developed in the project are expected to fit the need.

In addition to its completeness and updated information, the BIM data should also be interoperable with BES in an easy and automatic manner without a need for the repetition of the exhaustive input data. BIM4EEB tools and methods were developed to also fulfil this need, which is not always possible with the existing tools. There is often a need for different levels of manual interventions.

The model for the BIM Assisted Scenario Simulator tool, called BIMeaser, is presented and the workflow of its use is defined. The most important finding related to the renovation process is the importance of the need for the performance-based building design in the early design phases, where the most important decisions are made according to costs and performance as shown in Fig. 1.3.

The data definitions related to the commercial energy simulation tool IDA Indoor Climate and Energy (IDA-ICE) by the Swedish company Equa Simulation AB open a landscape of huge set of details to be defined. This huge set (more than 1000 parameters) was narrowed according to the findings listing the most important input variables related to the impact on the indoor climate and energy, but still was a big challenge in the project.

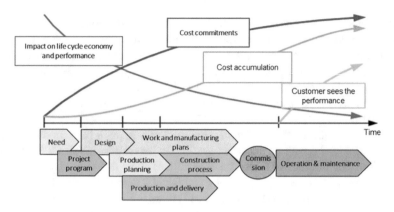

Fig. 1.3 Committed costs and possibilities related to energy performance decay, adapted from (Pietiläinen et al. 2007)

The second finding derived from tool definition was the need for the enhanced collaborative design work. The indoor climate and energy design is a multi-domain challenge, and it should always be considered as a teamwork. All tools around the future building renovation design should comply and support this approach. The performance-based building design process assumes that design selections are validated against the Owners Project Requirements (OPRs) in each design stage before moving to a following design stage. The design team will handle the detailed technical selections affecting to the OPRs using the tool. The management of the consistency of the detailed technical parameters against the OPRs has traditionally been handled manually between design domains resulting to the design errors or time-consuming updates, whenever the technical detail has changed. The BIM Assisted Scenario Simulator will tackle the complexity of the energy and indoor climate design by:

1. speeding up the decision making
2. enhancing the collaboration between the design domains
3. enabling the cross-domain transparency of the technical details in the design team
4. resulting better indoor climate and energy selections in the renovation.

The implemented BIM Assisted Scenario Simulator, called BIMeaser, is presented in Sect. 5.2.

1.3 Information Requirements for Construction Companies in Renovation Interventions

In the last years some studies have been developed regarding the application of BIM in the construction enterprises, but research on BIM adoption in small and medium-sized enterprises (SMEs) have remained an under-represented area (Hosseini et al.,

2016; Li et al., 2019). The research on SMEs it is of fundamental importance because SMEs have greatly contributed to the economic development of regions or countries and, currently, guidance to assist SMEs to make an informed decision about BIM adoption are lacking (Lam et al., 2017). It is necessary to understand the main challenges delaying the adoption of BIM in SMEs and to consider corresponding strategies that can be applied to obtain further understanding of BIM in SMEs (Li et al., 2019).

Focusing on the small and medium enterprises, the main barriers derived from the risks associated with an uncertain Return On Investment (ROI) for BIM implementation. The findings of (Hosseini et al., 2016) show that currently around 42% of Australian SMEs use BIM in Level 1 and Level 2 with only around 5% that have tried Level 3. According to (Lam et al., 2017) there is evidence to suggest that small and medium sized enterprises are currently losing out in winning publicly funded projects. (Hosseini et al., 2016) affirm that the findings of their study show that around 42% of SMEs have been engaged in BIM, instead of the 25% provided by the study by (Gerrard et al., 2010). This gap indicates how fast-moving BIM is within the Australian construction industry, but the findings also indicate that the immaturity of BIM implementation is still a problem within Australian market because only the 5% of SMEs had used Level 3 and 8% had utilized Level 2 BIM on their projects. According to (Li et al., 2019), based on previous studies and interviews, in SMEs six challenges exist: (1) SMEs are short on resources, (2) collaboration challenges, (3) lack of BIM awareness, (4) legal disputes and uncertainties in policies, (5) difficulties in meeting SMEs' needs, and (6) concerns about data and information. (Bosch-Sijtsema et al., 2017) assert that the results from their study on SMEs indicated that more than half of medium-sized contractors in the sample used BIM in some projects and the main obstacles for BIM implementation is related to the lack of normative pressure. (Tranchant et al., 2017) carried out an immersive 3D projection for the simulation of the Ajaccio hospital in France. According to (Tranchant et al. 2017), the heart of the economy of France is that of SMEs, which represents 96% of companies. As a result, the innovation of the BIM is becoming a necessity for SMEs, even if they do not practice public contracts. (Malacarne et al., 2018) affirm that the BIM Pitch concept that they developed in their research is unsuitability for SMEs since it is highly time consuming to develop as additional model, as consequence further study are required.

The framework on the recent studies on SMEs underlines that they are limited to few countries. As consequence, additional research is needed to understand the needs of the industry and existing barriers in digitalization and BIM use in AECOO sector.

The exchange of information between the three main parties of a project is a common problem in construction sector.

1. Communication and information exchange between the client and the contractor (under the word "client" are also considered the design teams)
2. Information exchange from the contractor to the supplier (under the word "supplier" is also considered the subcontractor)

3.   Information exchange from the supplier to the contractor

The traditional information exchange is done by paper or digital format, using graphics, pictures, documents, and static media documents. This type of exchange of information has several limitations that can be improved utilizing a more digitization method such as BIM.

The BIM method includes the use of information models to replace documents that can be digitalized, and it optimizes the transmission of data, promoting interoperability between the parties involved in the project.

However, the geometric representation is not the only purpose of a BIM model; in fact, this type of approach is used with the addition of data and information characterizing the model itself. From the informative model, it is possible to extrapolate, in addition to the geometric visualization, also data of an informative nature and the latter can be contained in documents linked to the model, such as quantities estimation, specifications or reports, or within the objects constituting the model itself.

Commonly, information exchange issues are related to three main parties involved in a project, namely client, contractors and supply chain as following explained based on (Franchini, 2018).

1.3.1   From the Client to the Company

At the base of the process, there is the need of the client to have the work carried out. The client, therefore, represents the main Fig of every transaction. To this end, the company should clearly understand what the client's objectives are.

To clarify these intentions, the client trusts the elaboration of the project to a team that draws up part of the documents that must be supplied to the companies.

In a frequent scenario, once they have received the material, they must analyse the information and data received from the design team and verify the completeness. If the papers are inadequate, the responsibility of filling these gaps will nevertheless remain with the company, to draw up a complete construction project.

Often, for example, the executive project is incomplete, leading to a loss of time and resources for the company.

At the end of the process, the company should deliver to the client the complete As Built, including all the information useful for the maintenance of the work. Currently many companies have an ERP system (such as SAP), where part of the As-Built information is collected. However, many documents remain stored in other location in the company there is the need to transfer all this inputs to the client and the end of the project.

1.3.2   From the Company to the Supplier

This exchange of information happens to present an offer.

The traditional exchange of information between the company and the supplier happens by paper or digital support, exploiting the use of graphic, documentary, and static multimedia documents. This type of information exchange is usually successful but has several limitations, including:

- The documents or files exchanged are not self-explanatory. Therefore, the meaning of the message may be subject to personal interpretation.
- The exchange of information through this type of support forces the subjects involved to operate in a non-optimal way.

This way often leads to the loss of time and valuable information, and the waste of a large amount of paper.

Even the world of construction companies is getting closer to a more complete digitization; an act that regards information exchange very closely. However, it is important to emphasize that this step forward by companies lies not only in the digitization of their process, but in the use of information models to replace the documents which can be either digital or not. All this makes it possible to optimize the transmission of data, promoting interoperability between subjects, limited by having to know how to use the tool.

If the company was digitized, in the BIM sense, the transfer of information to suppliers would be difficult, if not impossible. The supplier would not be able to open and exploit the information models received, because he is used to using antiquated methods (paper) and/or non-standardised practices. In fact, a supplier may deliver products to hundreds of companies and each of them may have different practices.

1.3.3   From the Supplier to the Company

This exchange of information happens to allow the company to present an offer; to realize the as built, and the maintenance documentation.

This passage of information is significant at different stages of the process starting from the pure design phase, in which the company must take care to carry out the constructive translation of the client's requests, up to the preparation of the maintenance plan that will be used by the final users.

The necessary inputs can be multiple, such as:

1. the date of installation of a material or the date of installation of a product,
2. information in relation to the acceptance of the material,
3. the possibility of replacing a product,
4. the supplier not digitized in the BIM sense, in the same way as in the previous case, will not be able to respond to the needs and requests of the company due to the lack of knowledge of the information method adopted.

There are two possibilities to solve this issue:

1. translating the informative models back into elaborates so that the supplier can analyse the documentation and work on this to respond to the business needs,
2. developing an intelligent method so that the supplier, despite not knowing the BIM platforms and software, is able to exploit these intuitive and practical methods to meet the needs of the company, together with those of the client.

The first point is feasible but would lead to a waste of time and resources that could instead be addressed elsewhere. It is necessary to carry out an accurate investigation to try to better understand the needs of the companies, in such a way as to be able to obtain results that are practically exploitable and standardized.

1.3.4 Attributes and Requirements

To explore solutions related to the above-mentioned information exchange issues, it is necessary to know some information regarding the companies, such as:

- What information does the supplier need to provide for the preparation of an offer?
- Which documents are useful for the preparation of an offer besides the specifications and the calculation?
- What are the necessary inputs to the model for the preparation of an As Built or a maintenance plan?
- Which of this information is essential to be digitized in order to optimize the process?
- Which of this information should be given exclusively by the supplier or site manager?
- What are the attributes that are needed to characterize an object?

To find out what are the attributes three types of objects are considered, representing the category to which they belong, such as:

- A construction work characterized by geometry without the use of industrial products.
- An industrial product without geometry in BIM (e.g., brick) with geometry defined by the construction work (e.g., wall).
- An industrial product geometry both in BIM and in the real world (e.g., air handler).

The first step is to create a basic list of attributes, useful for the company to characterize the object.

It is important that this list can be implemented to be able to add both specific needs and attributes, expressly requested by the client or designer, and new attributes considered essential by the supplier (Table 1.1).

- Is each proposed attribute necessary?

- Is it essential to identify others, and if so, which ones?
- Which of these are needed in the offer phase, such as for the preparation of an As Built, and what for a maintenance plan?
- Which of these attributes is essential to be digitized to optimize the process?
- Which of these attributes should be given exclusively by the supplier or by the site manager?

1.4 Definition of Owner and Inhabitants' Needs and Requirements in Renovation Interventions

An important and difficult step in designing a software product is determining what the end-user expectations of the software will be, such as what activities the user will be able to perform once the platform is operational. This is because the users (especially non expert's users) are often not able to communicate entirely their needs, or the information they provide may be incomplete, inaccurate and/or self-conflicting. On the other hand, nowadays there is the increasing need of the active enrolment of end users towards the provision of personalized services to the potential customers.

The scope of this document is to blueprint a methodological approach and a user requirements extraction process for the building owners' and inhabitants' as main stakeholders in a building renovation and management project. In general, by engaging the owners and inhabitants of a building, enables them to participate as active players in the renovation process, which in turn increases their awareness about their influence on the way a building function. As such, their input during the whole process ensures that their requirements and needs in relation to the renovation's outcome are considered; additionally, it increases the chances of a successful renovation design outcome achieved.

Towards this direction, a detailed methodology is established at the early phase to ensure the proper extraction of end-user's feedback. The definition of the target groups is a primary step along with the definition of the main use cases associated with the roles. Then, to extract the relevant requirements, dedicated to the building owners and inhabitants' questionnaires are structured, while semi-structured interviews (to be addressed by the building representative) were defined, towards co-creating a shared value and directly addressing the owners and inhabitants needs. More details about the methodological framework as well as the results from the analysis are presented in the following.

As shown in figure below, the proposed methodology is composed by three steps through which the requirements coming from the building owners and inhabitants can be captured.

The different steps shown in Fig. 1.4 are detailed in the following text.

Table 1.1 Description of informative attributes

Attribute	Description
Accessibility	Accessibility performance (ability to solve problems in this sense)
Type	Flexibility of the object
Category	Coding indicating their classification (e.g., Uniclass2015)
Code performance	Category requirements
Colour	Primary colour or characterizing the product
Component	Optional components (parts, characteristics, and finishes)
Description	Description of the object that clarifies the design intent
Useful life	Life expectancy of the object (typically indicated in years)
Use	Typical service use
Characteristics	Other essential features related to product specifications
Finish	Main finish
Category	Standard category
Manufacturer	Useful manufacturer's contact
Material	Main material
Serial Number	The code assigned to the object by the manufacturer
Name of the model	Name of the object used in the model
Name	Alphanumeric code that uniquely identifies the object (the code must first report the type of product)
Nominal height	Size typically measured vertically
Nominal length	Size typically measured horizontally
Nominal weight	Object weight
Cost of replacement	Indicative cost for replacement
Form	Main or characteristic form
Dimension	Main or characteristic dimension
Sustainability	Description of sustainability requirements
Warranty	Warranty description and possible exclusions
Duration of warranty (installation)	An indication of the duration of warranty of the installation
Duration of warranty (components)	An indication of the duration of warranty of the components
Duration of warranty (object)	Duration of warranty of the object (normally in years)
Guarantor warranty	Contact details of the guarantor for the installation

(continued)

Table 1.1 (continued)

Attribute	Description
Guarantor of the guarantee (component)	Contact details of the guarantor for the components
Specific identification	Identification of a specific activity to distinguish it from the others
Bar code	Bar code or RFID for unique identification of the object
Date of installation	Date the object was installed
Serial number	An indication of the serial number
Tag	Tag indication
Warranty start date	Date the warranty begins

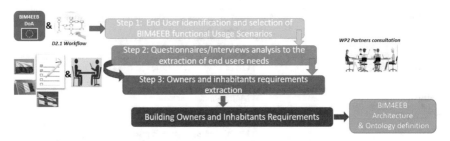

Fig. 1.4 Steps of the followed workflow

Step 1: Selection of end users usage scenarios

The starting point for this analysis, is the identification of the end-users (namely the building owners and inhabitants) business needs and priorities, following consultation with building occupants. A non-exhaustive list of usage scenarios is defined for analysis in the project:

- US-01: Establishment of a comfort and IAQ preserving framework for inhabitants during/post renovation period.
- US-02: A continuous interaction framework for comfort status monitoring and report for inhabitants
- US-03: An alerts and notification framework during the renovation process for owners and inhabitants.
- US-04: Management and control of renovation interventions for owners.
- US-05: Control of working times and economic sustainability for owners - tracking tool for renovation operations for owners

The definition of the different Usage Scenarios highlights the key differentiation among the stakeholders (owners versus. inhabitants).

Step 2: Questionnaires/Interviews analysis to the extraction of the end-users needs

Following the definition of the Usage Scenarios, a set of questionnaires were prepared to address the needs of the building owners and inhabitants (as end-users of the project). The scope was to engage these end-users in the project's activities and further retrieve their valuable feedback towards the extraction of their needs and requirements related to building renovation and management activities. The intended type of information to be gathered from the end users is classified into the following groups:

- Profile information: such as profile data, household composition and building characteristics. This information is useful for segmentation and statistical analysis.
- General knowledge and familiarity with concepts such as BIM (Building Information Modelling), BACS (Building Automation and Control systems), etc., and questions trying to capture the current knowledge of the Owners and Inhabitants, regarding how their building/premises operates.
- Owners & Inhabitants required tools and attitude towards participation in the usage scenarios. This is the targeted information aiming to extract the priorities of the users related to the different usage scenarios defined.

Further to the use of questionnaires for extracting the owners' and inhabitants' needs and requirements, the building site representative partners were also engaged in semi-structured interviews to extract further pilot specific requirements. The use of these semi-structured interviews allowed new ideas/concepts to be examined, as well as further technical requirements to be identified regarding ethical/legal barriers imposed at the different pilot sites. More specifically, the goal of these semi-structured interviews was threefold:

- To shape the final list of Usage Scenarios addressing the owners and inhabitants needs.
- To cross check the results from questionnaire analysis towards the extraction of the final list of owners and inhabitants' requirements.
- To gather any regulatory and legal requirements as part of the overall requirements definition phase at the different pilot sites.

A total number of 102 questionnaires were responded by building owners and inhabitants—45 for building owners & 57 for building inhabitants. Also, a limited number of semi-structured interviews was answered by the pilot representatives (Italy, Poland, and Finland) to express their needs and main expectations from the project.

Step 3: Owners and Inhabitants requirements extraction

Finally, in step 3 we consider the questionnaires/interviews analysis results which are transformed to building owners and inhabitants requirements. A total number of 97 end-user's requirements were extracted incorporating also some technical constrains. The list of the extracted requirements derived from the active participation of the pilot sites. These are further complemented by additional requirements mainly related to ethical and any legislation constrains imposed at the pilot regions of the project.

In total, 30 pilot specific requirements are to be considered at the instantiation/configuration of the framework at the different pilot sites. Further to the above work and the elicitation of the non-exhaustive list the requirements, we categorized them considering the hierarchy and the project's objectives. The prioritization of the end-user's requirements is also a critical task performed to elucidate the requirements to be considered on the analysis. The different layers of priority are:

- High: Requirements in this category as defined are a key innovation of the project. These requirements are essential to achieve the goals of the project and fulfil the end-users' needs.
- Medium: These requirements are necessary or very helpful to set the application prototypes, but not crucial one for the whole system operation.
- Low: Requirements in this class are not necessary for the BIM4EEB system. However, they may be considering as important for the fine-tuned operation of the system examined.

Overall, the outcome of the work is (a) the definition of the relevant for the building owners/inhabitant's usage scenarios to further enable the (b) extraction of the user specific requirements for this target group of the BIM4EEB project. The extraction of the requirements will further facilitate the design and development of the different software implementations targeting building occupants: inhabitants and owners.

References

BIM4EEB D2.1 Definition of relevant activities and involved stakeholders in actual and efficient renovation processes (2019)

BIM4EEB D2.2 List of Designers' Needs and Requirements for BIM-Based Renovation Processes (2019)

Bosch-Sijtsema P, Isaksson, A., Lennartsson M, Linderoth H (2017) Barriers and Facilitators for BIM Use among Swedish Medium-Sized Contractors—'We Wait until Someone Tells Us to Use It' (March 22, 2017). Bosch-Sijtsema P, Isaksson A, Lennartsson M, Linderott, HCJ (2017) Barriers and facilitators for BIM use among Swedish medium-sized contractors—"We wait until someone tells us to use it". Vis Eng 5(3):1–12

CCC California Commissioning Collaborative, California Commissioning Guide: Existing Buildings (2006) https://cxwiki.dk/files/stream/public/CA_Commissioning_Guide_Existing.pdf. Accessed 07.03.2022

EN 16310:2013—Engineering services—Terminology to describe engineering services for buildings, infrastructure and industrial facilities

Franchini C (2018) Scambio informativo nei processi edilizi: ottimizzazione della comunicazione tra imprese e fornitori—definizione dello strumento di analisi. (master thesis, supervisor Alberto Pavan). Politecnico di Milano, Italy

GBPN Global Buildings Performance Network, What is a deep renovation definition?, (2013) https://www.gbpn.org/wp-content/uploads/2021/06/08.DR_TechRep.low_.pdf. Accessed 07.03.20

Gerrard A, Zuo J, Zillante G, Skitmore M (2010). Building Information Modeling in the Australian Architecture Engineering and Construction Industry. In J Underwood, U Isikdag (Ed.), Handbook of Research on Building Information Modeling and Construction Informatics. Concepts and Technologies (pp. 521-545). IGI Global. https://doi.org/10.4018/978-1-60566-928-1

Hosseini MR, Banihashemi S, Chileshe N, Namzadi MO, Udeaja CE, Rameezdeen R, McCuen T (2016) BIM adoption within Australian small and medium-sized enterprises (SMEs): an innovation diffusion model. Construction Economics Building 16(3):71–86

ISO 6707-1:2020 Buildings and civil engineering works—Vocabulary—Part 1: General terms

ISO 6707-2:2017 Buildings and civil engineering works—Vocabulary—Part 2: Contract and communication terms

Lam TT, Mahdjoubi L, Mason J (2017) A framework to assist in the analysis of risks and rewards of adopting BIM for SMEs in the UK. J Civ Eng Manag 23(6):740–752

Li P, Zheng S, Si H, Xu K (2019) Critical challenges for BIM adoption in small and medium-sized enterprises: evidence from China. Adv Civ Eng

Malacarne G, Toller G, Marcher C, Riedl M, Matt DT (2018) Investigating benefits and criticisms of bim for construction scheduling in SMEs: an Italian case study. Int J Sustain Dev Plan 13(1):139–150

Pietiläinen J, Kauppinen T, Kovanen K, Nykänen V, Nyman M, Paiho S, Peltonen J, Pihala H, Kalema T, Keränen H. ToVa-käsikirja. Rakennuksen toimivuuden varmistaminen energiate-hokkuuden ja sisäilmaston kannalta [Guidebook for life-cycle commissioning of buildings energy efficiency and indoor climate]. Espoo 2007. VTT Tiedotteita—Research Notes 2413. 173 p. + appendices 56 p. https://www.vtt.fi/inf/pdf/tiedotteet/2007/T2413.pdf. Accessed 07.03.2022

RIBA (2020) Plan of Work 2020 Overview. Available at: https://riba-prd-assets.azureedge.net/-/media/Files/Resources/2020RIBAPlanofWorkoverviewpdf(1).pdf?rev=c968ceebdcce42e09c90aaa77cff7044;hash=CAD5C5236AA1C1869F37FB92364FFB71

Tranchant A, Beladjine D, Beddiar K (2017) BIM in French SMES: from innovation to necessity. WIT Trans Built Environ 169:135–142

Chapter 2
Linked Data and Ontologies for Semantic Interoperability

Karsten Menzel, Seppo Törmä, Kiviniemi Markku, Kostas Tsatsakis, Andriy Hryshchenko, and Meherun Nesa Lucky

Abstract The purpose of Linked Data and ontologies is to achieve semantic interoperability across different systems and agents. As a concrete implementation, Digital Construction Ontologies (DiCon) are meant to provide efficient information sharing and utilization between the tools in the BIM4EEB toolkit. This section explains the motivations, design principles, and organization of DiCon ontologies.

Keywords Semantic interoperability · Linked data · Ontology

2.1 The Role of Linked Data and Ontologies

Construction and renovation industry is nowadays filled with digital solutions to support numerous tasks in design, planning, quantity estimation, procurement, logistics, site management, quality inspections, sensor data gathering, and so on. However, due to their specific origins and history, different solutions manage and represent data in an incompatible manner. The result is a fragmented systems landscape and the

K. Menzel
Technische Universität Dresden (TUD), Dresden, Germany

S. Törmä (✉)
VisuaLynk Oy, Espoo, Finland
e-mail: seppo.torma@visualynk.com

K. Markku
VTT Technical Research Centre of Finland Ltd., Espoo, Finland

K. Tsatsakis
SUITE5 Data Intelligence Solutions Limited, London, UK

A. Hryshchenko
University College Cork (UCC), Cork, Ireland

M. N. Lucky
Department ABC—Architecture, Build Environment and Construction Engineering, Politecnico di Milano, Milan, Italy

© The Author(s) 2022
B. Daniotti et al. (eds.), *Innovative Tools and Methods Using BIM for an Efficient Renovation in Buildings*, SpringerBriefs in Applied Sciences and Technology,
https://doi.org/10.1007/978-3-031-04670-4_2

consequent need for repeated manual work to transfer information from one system to another or for costly integrations between pairs of systems.

Linked Data and ontologies provide a generic solution that enables systems to interoperate; that is, share information and act on the information shared. The need for manual information exchange is reduced or even removed since the interaction between systems can be automatized.

The following levels of interoperability can be identified (Singh and Huhns 2005):

1. *Technical interoperability*: At the lowest level there must be a *connection* between the systems, and an *interface* through which bits and bytes can be transferred from one system to another. This can be achieved if the systems are connected to a common communication network and if they have application programming interfaces.

2. S*yntactical interoperability*: There should be common understanding regarding the format of transferred data, so that the recipient can parse the structure of the data. This can be solved by using standard data formats (e.g., JSON, XML, CSV, or STEP Physical File Format).

3. *Semantic interoperability*: The terms used in the data (types of entities, their properties, datatypes, and identifiers) should be understood in a sufficiently similar manner by the systems to make their practical operations successful. The solutions for semantic interoperability are still in development with approaches based on standards, ontologies, wrappers, and mediators (Abukwaik, 2014). In the construction domain also classification systems can address some aspects of semantic interoperability.

4. *Pragmatic interoperability*: There are several issues related to processes, security, adaptation to dynamic changes, organizational arrangements, and even legal considerations where the systems may need to be in an agreement to successfully work together. Occasionally some of these aspects are regarded as additional layers of interoperability (EIF, 2017) (Abukwaik, 2014).

The solutions to tackle technical and syntactic interoperability, at least in ordinary domains, are already available, and to achieve interoperation at those levels is generally a matter of willingness and effort of implementation. Semantic interoperation is still an area of active development, with promising technologies and initial results. However, even after the problems of semantic interoperability have been tackled, challenges of the pragmatic level remain to be solved to make systems work seamlessly together.

In the case of the BIM4EEB toolkit, the Digital Construction Ontologies and the Linked Data principles take care of the semantic interoperability but as of now the solutions for the problems arising at the pragmatic level must still be separately agreed among the tools.

2.2 The Need to Supplement BIM with Ontologies

While in the future the design in renovation projects will be based on building information modelling (BIM), the information contained in the models will not be sufficient to comprehensive planning and modelling of different aspects of a renovation. Rather, BIM models will be related to multiple other areas as shown in Fig. 2.1: built assets, agents, legal roles, products and materials, classifications, information entities, plans, activities, sensor data, and so on. The entities in these areas need to be linked with the entities in BIM models but the different origin, lifecycle, and rate of change of the data in those domains imply that they are best managed separately from BIM models. That is, interlinked but separately managed.

There are several areas where comprehensive renovation management requires supplementary representational agreements:

1. *Occupancy and ownership.* In renovations, information about ownership, occupancy, and use of spaces of the building to be renovated require the concepts for built assets, including their division into individual residential and non-residential units in buildings. This is used in the early stages of renovation planning to organise the information about occupancy profiles, and requirements and preferences of stakeholders. In the execution phase it is used to coordinate the renovation activities with the stakeholders.

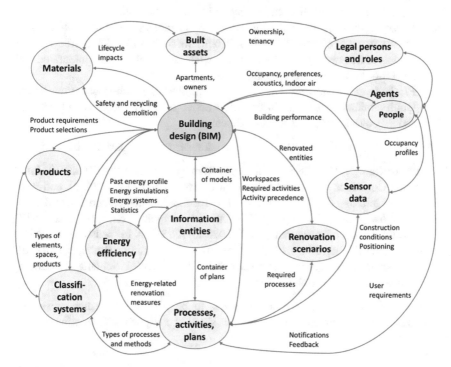

Fig. 2.1 Examples of domains relevant for renovation management

2. *Information models of assets and projects.* All construction projects are characterized by numerous information entities: BIM models, breakdown structures, quantity take-offs, operational plans (e.g., master plans, lookahead plans, week plans), drawings, and so on. The processes of creation and versioning of these needs to be supported. A specific concern is the linking of appropriate parts of project information model to the asset information model of the building.

3. *Renovation scenarios and renovation measures.* In the early planning, different renovation scenarios can be explored, each consisting of a set of renovation measures. Renovation measures are activities that are often repeated to different parts of a building identifiable from BIM models: for example, for each bathroom, window, part of facade, and so on.

4. *Construction management.* The concepts of lean construction can be modelled as activity flows, that is, relations of activities to building objects (from BIM), locations (partly from BIM), labour crew, equipment, materials, information, and external conditions (from IoT). These are needed to support methods such as Location-Based Management and Takt Time Planning (Seppänen, 2014), or practices like the Last Planner™ (Ballard, 2000).

5. *Sensor data.* Sensor observations can be organised by relating them to objects in the BIM models, such as spaces or building elements, the resources at the construction site (people, equipment, materials), or entities in the supply chain (shipments, products). Through the connections of activities to these entities, the progress of execution can be derived from sensor data.

6. *Energy efficiency.* An important contemporary motivation of renovations is the improvement of energy efficiency of a building over its remaining lifecycle. The models of life-cycle assessment and life cycle costing need to be integrated with BIM models and projects management.

7. *Products and materials.* The ability to promote low-carbon construction and facility maintenance, the information about materials, layers, and products should be represented in a detailed manner.

The representation of these areas has been addressed at least partially by many

- *standards*: e.g., BIM-based Information Management (ISO19650, 2018), Information Container for Linked Document Delivery (ISO 21597, 2020), and Basic Formal Ontology (ISO/IEC 21838, 2021) for fundamental categories, and
- *established ontologies*: e.g., OWL-Time for temporal concepts (Cox, 2017), QUDT for units of measure (QUDT, 2022), SSN/SOSA for sensor observations and systems (Haller, 2019), PROV-O for origin of information (Lebo, 2013), FIBO for financial and legal concepts (Bennet, 2013), or Brick Schema for building automation systems (Balaji, 2016).

The previous work on ontologies needs to be considered in any subsequent ontology development, and new ontologies should be aligned with existing ones.

2.3   Tool Interoperation Using Ontologies

The applications that utilize ontologies can originally be standalone tools. It is expected that future solutions for renovation management are likely to be implemented as systems of systems (Maier, 1998). There will be independently developed and used systems serving their specific purposes that will produce and consume information. The overall information system environment in any specific renovation project will be built from the existing systems of the parties involved. The essential goal is to have interoperability between different systems because it allows to connect them together to serve the goals of the project.

The constituent systems must relate to each other for the duration of a renovation project to act in complementary roles in the project. Detailed data remains inside each system, but the systems can share the relevant part of their data with others. To achieve interoperability, the constituent systems must have interfaces for the required interactions and semantic interoperability relevant to the renovation domain.

Figure 2.2 illustrates the role of the ontologies with a high-level architectural diagram as realized in the BIM4EEB project. There is a set of renovation tools that interoperate by sharing data through a common data sharing platform. The tools do not interact with each other directly; all interactions happen through shared data.

Data sharing is based on the Linked Data principles and the use of ontologies that enable the semantic interoperability among tools. Both the tools and the data sharing platform access the ontologies either by downloading whole ontology modules or through URI lookup of individual terms. Downloading of whole ontologies is needed for performance reasons and to support reasoning functionalities.

The data sharing platform stores data in files and in databases. There is both a relational database (SQL) and a graph database (RDF) between which the data contents are synchronized.

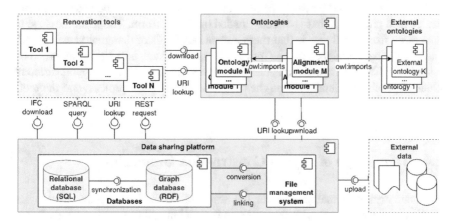

Fig. 2.2 An architectural overview ontology-based toolkit of BIM4EEB

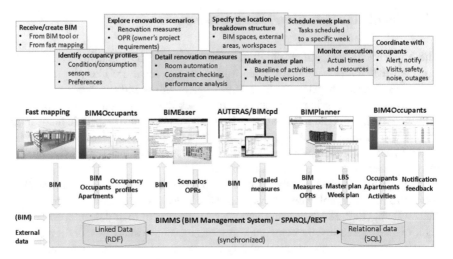

Fig. 2.3 High-level example of data sharing between tools in a workflow

The tools can access the data sharing platform in several different ways: by downloading IFC files and through SPAQRL queries, URI lookup and ordinary REST interface. Finally, the ontology modules are aligned with external ontologies through the alignment modules that import both the external and internal ontologies and connect the terms across them with alignment axioms.

A high-level example of the kind of data sharing that can happen between different tools in the workflow of BIM4EEB toolkit is shown in Fig. 2.3. The tools are shown in the middle level of the figure. The tasks that are performed with the aid of the tools are indicated at the top of the figure. At the bottom is the data sharing platform. The arrows between the tools and the data sharing platform describe the kinds of data that is exchanged between the tools and platform. It should be noted that there can be also other data that is directly loaded on the platform. The main message to take away from the figure is that there are various kinds of data that is exchanged, and that the data produced by one tool can later be utilized, enriched or modified by other tools. The interoperation is based on shared data, and it can support different kinds of processes, depending on the companies involved in projects and the tools utilized by them.

2.4 The Design of Digital Construction Ontologies

The main principles that have driven the design of Digital Construction Ontologies (DiCon) are compliance with relevant standards, compatibility with a standard top-level ontology to provide fundamental categories and therefore increase conceptual integration, support for information evolution which is a crucial phenomenon in renovation and construction projects, the flexible mechanism to connect the entities with different identifier schemes and classification systems.

2.4.1 Standards Compliance

An important objective in the design of Digital Construction Ontologies was to support relevant standards in the AEC domain. Most important standard is IFC (Industry Foundation Classes) (ISO16739, 2018) that has also been converted into an ontology ifcOWL (Pauwels, 2016). The BIM models created by designers can be exported into IFC, and the IFC model contains all spatial and physical entities as well as building systems of a buildings. Another important standard is ISO 19650 that specifies the principles of BIM-based information management. DiCon provides a formalization of the central concepts of ISO 19650 standard (ISO19650, 2018) ranging from built assets, appointments and project teams to information models consisting of a federation of information containers. Finally, the information-related concepts of DiCon are compatible with the Container ontology of Information Container for Linked Document Delivery (ISO 21597, 2020) that enables the exchange of packages of interlinked datasets across parties.

2.4.2 Top-level Ontology for Fundamental Categories

Since DiCon covers a wide range of concepts—physical entities, spatial entities, information entities, events, processes, properties, roles, capabilities, etc.—there is a need for generalized concepts or higher-level categories as one method to interrelate different domains of information. Instead of ad hoc generalizations, a standard and widely used top-level ontology Basic Formal Ontology (ISO/IEC 21838, 2021) is used as the source of fundamental categories. BFO is a compact ontology based on ontological realism—according to which ontology represents entities in the world—that better promotes interoperability than ontologies that allow cultural or mental concepts with more diverse interpretations.

2.4.3 Support for Information Evolution

In renovation and construction projects it is crucial to support the evolution of information. Information will be accumulated, refined, and changed during the project execution. DiCon supports the evolution of information with two different mechanisms. Firstly, objectification of properties is used for the small-scale evolution, e.g., for values changing over time because of sensor observations. Objectification also enables the representation of other metadata about properties: quantity kinds, units of measure, and property definition entities. Secondly, the large-scale evolution is based on the ISO 19650 information containers. This enables the storage and management of alternative renovation scenarios, different versions of BIM models, releases of operational plans, and entity organizations such as breakdown structures.

2.4.4 Flexible Labelling of Entities

All DiCon entities can be labelled with identifiers and classifications. Every independent entity has a URI as its main and retrievable identifier but, in addition, the entity can have additional identifiers in different information systems and other scopes. Identifiers can be globally unambiguous such as a GUID, or locally unambiguous, such as a room number in a building (where the building defines the scope of the identifier). Likewise, since different classification systems in the construction sector are used in different geographical regions and organizations, DiCon supports the flexible labelling of entities with categories based on the needs of a project. The goal of DiCon is to avoid duplicating information that is already in classification systems; after all, classifications are complementary to ontologies in the sense that they capture the variety in the domain with an extensive set of detailed classes (possibly with class-specific attributes), while ontologies represent the complexity in it with a compact set of essential classes, each with a distinct set of constraints and relations to other classes.

2.4.5 Modularization and Alignment Using Vertical/horizontal Segmentation

DiCon is modularized using the vertical and horizontal segmentation approach of the Semantic Sensor Network Ontology (Haller, 2019). In the *vertical dimension*, a new module imports the previous one and deepens the representation of the underlying domain by defining additional subclasses, properties, restrictions, or alignments (therefore, alignment is always vertical segmentation). In the *horizontal dimension*, a new module broadens the domain by defining classes complementary to the previous ontology as well as properties to connect them to the previous concepts.

In the vertical dimension, the new module should support selected use cases better, having perhaps a narrower user base when compared to the previous module, while horizontal segmentation should extend the set of supported use cases, therefore broadening the potential user base.

In accordance with vertical segmentation, the approach to integrate DiCon ontologies with external ontologies is based on the use of alignment modules. It should be noted that there are several alternative ways to promote integration: (1) direct imports of external ontologies, (2) references to terms of external ontologies without importing them, and (3) utilization of SHACL to create application profiles as graph templates that incorporate terms from several ontologies. The resulting integration is different in each case; for instance, the consistency checking with ontology reasoning works in a different manner.

In the alignment approach the ontology modules to be aligned do not reference each other at all. Instead, a separate alignment module first imports the ontologies to be aligned and, second, provides additional alignment axioms that establish relationships between the terms in the imported ontologies. The alignment module can then be checked for consistency to reveal whether the alignments are compatible with the definition of the terms in the aligned ontologies.

According to the alignment approach all the references from DiCon ontologies to classes and properties defined in external ontologies are made explicit in the alignment modules; that is, the DiCon ontologies do not refer to any external classes or properties. However, it is still possible that they have references to individuals defined in external vocabularies. This applies especially to individuals representing quantity kinds and units of measure.

2.5   Definition of DiCon

The overall DiCon ontology is divided into a set of modules shown in Fig. 2.4. The arrows indicate the import relationships (owl:imports) between the modules.

Contexts—The DiCon Contexts (dicc) ontology defines the basic concepts to manage data in different alternative realms, such as planned or actual data, models at different levels of detail, different renovation scenarios, versions of construction plans, and so on. The main concept is dicc:Context, each of whose data is contained in its own named graph of an RDF dataset.

Variables—The DiCon Variables (dicv) defines an enriched representation of the properties. It uses the objectification of properties as the basic mechanism. The properties of entities can be associated with additional objects to represent quantity types or units. Moreover, the value can be represented with property states that enable the values at different time points or levels of detail. The ontology is aligned with

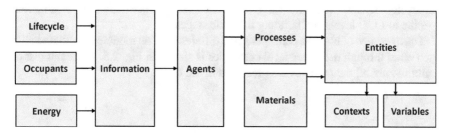

Fig. 2.4   DiCon ontology modules and their import relations (owl:imports)

QUDT, SOSA, and Saref Core. It allows to associate, for instance, series of sensor values to the properties of entities.

Entities—DiCon Entities (dice) defines the basic entities of renovation and construction processes, ranging from building objects, real estate, devices, equipment, locations, roles, and capabilities to spatial regions and time intervals. The ontology observes the general structure of the Basic Formal Ontology (BFO) and is aligned with it.

Processes—DiCon Processes (dicp) defines the concepts for activities (intentional processes with start and end), activity flows and resources. An activity can have any number of sub activity layers, and it can represent anything from projects to tasks. The activity flows are relations that activity can have to different other entities: the object to transform, agents, locations, equipment, material batches, information entities, and to external conditions such as temperature or humidity. Resource is treated as a role that any entity can have with respect to an activity and activity may require different kinds of resources to appear in different activity flows.

Agents—DiCon Agents (dica) represents the concepts for actors and stakeholders, persons, organizations, and legal persons, as well as the basic organization concepts from ISO19650. The legal persons can have specific roles with respect to assets, for instance, as an owner or tenant.

Information—DiCon Information (dici) defines concepts of information content entities such as information containers, information models, designs, plans, messages, and so on. The concept of a project information model is defined based on the description in ISO19650.

Materials—DiCon Materials (dicm) ontology defines the layer structures of building objects, and different kinds of properties of materials from carbon content to thermal and dynamic properties.

Energy—DiCon Energy (dices) defined the concepts for life cycle assessment, energy efficiency and energy systems, including the related information content entities, processes, and objects.

Occupancy—DiCon Occupancy (dicob) covers the occupancy profiles of inhabitants, and acoustic performance and indoor air quality of building units.

Lifecycle—DiCon Lifecycle (dicl) addresses the representation of information specific to LOD levels and building lifecycle stages.

The overview of how the concepts in the different DiCon modules are related with each other through the upper-level categories is shown in Fig. 2.5. In the interest of legibility, not all the modules are presented in the figure.

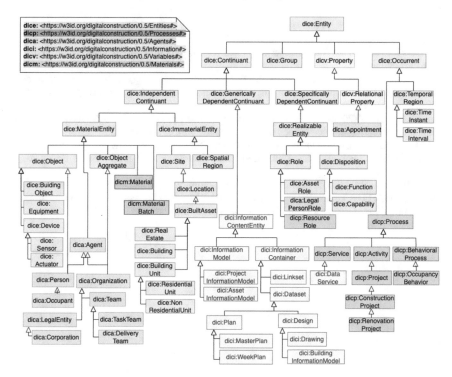

Fig. 2.5 Overview of central concepts of DiCon

References

Abukwaik H, Taibi D, Rombach D (2014) Interoperability-related architectural problems and solutions in information systems: a scoping study. In: Avgeriou P, Zdun U (eds) Software architecture. ECSA 2014. Lecture notes in computer science, vol 8627. Springer, Cham. https://doi.org/10.1007/978-3-319-09970-5_27

Ballard G (2000) The Last Planner System of Production Control, Doctoral dissertation, University of Birmingham

Balaji B, Bhattacharya A, Fierro G, Gao J, Gluck J, Hong D, Whitehouse K (2016) Brick: Towards a unified metadata schema for buildings. In Proceedings of the 3rd ACM International Conference on Systems for Energy-Efficient Built Environments (pp. 41–50)

Bennett M (2013) The financial industry business ontology: Best practice for big data. J Banking Regul, 14(3), 255–268

Cox S, Little C (2017) Time Ontology in OWL. W3C Recommendation

EIF (2017) New european interoperability framework, Promoting seamless services and data flows for European public administrations. Publications Office of the EU, Luxembourg

ISO/IEC 21838-1:2021 Information technology—Top-level ontologies (TLO)—Part 1: Requirements

ISO19650:2018 Organization and digitization of information about buildings and civil engineering works, including building information modelling (BIM)—Information management using building information modelling.

ISO16739-1:2018 Industry Foundation Classes (IFC) for data sharing in the construction and facility management industries—Part 1: Data schema

ISO 21597:2020 Information container for linked document delivery—Exchange specification—Part 1: Container

Lebo T, Sahoo S, McGuinness D (2013) PROV-O: The PROV Ontology, W3C Recommendation

Maier MW (1998) Architecting principles for systems-of-systems. Systems Engineering: J Int Counc Syst Eng 1(4), 267--284

Pauwels P, Walter T (2016) EXPRESS to OWL for construction industry: Towards a recommendable and usable ifcOWL ontology. Automation in Construction, 63, pp.100–133

QUDT; Quantities, Units, Dimensions and Types, DOI: 10.25504/FAIRsharing.d3pqw7, Last Edited: Friday, May 6th 2022, 12:03, Last Editor:delphinedauga,Last Accessed: Friday, May 6th 2022, 16:45

Singh MP, Huhns MN (2005) Service-oriented computing. Wiley, Chichester, pp 410–495

Seppänen O (2014) A comparison of Takt Time and LBMS planning methods. In Proc. 22nd Ann. Conf. of the Int'l Group for Lean Construction (pp. 23–27)

Chapter 3
Development of BIM Management System

Alberto Pavan, Vittorio Caffi, Alessandro Valra, Davide Madeddu, Diego Farina, Jacopo Chiappetti, and Claudio Mirarchi

Abstract With BS 1192:2007 and even more so with BS PAS 1199-2:2013 and 1192-3:2014, the concept of Common Data Environment (CDE) of the order (project, construction or management that is). Originating from a standard dedicated to design (BS 1192:2007) and although its concept has been extended to information management in general: Capex (strategy, project, construction: PAS 1192-2) and Opex (exercise: PAS 1192-3), the CDE, as it is understood today in common practice, is still very much affected by the original link with the design and the design phase (and in particular the design in the new building). All this according to an information flow that is still very linear and sequential: client, designer, builder, manager, user; more than circular, as the so-called BIM methodology would like. The risk, therefore, is that the commercial software market is affected by this CDE approach, which is also useful for the very rich real estate market of the emerging economies, neglecting the construction market of the more consolidated economies (Europe for before), very built up, and aimed more at housing quality, sustainability, reuse, and renewal of the existing heritage rather than the new one. It is consequently necessary to define new information flows and a new type of information management environment (CDE) for the phases of use, conservation, and renovation of buildings for the European market. The need arises for a specific BIM Management System (BIMMS, overcoming the classic CDE) for asset management and their enhancement that collects information from the buildings themselves and its users (Digital Twins, IoT, sensors, etc.). A new CDE / BIMMS that uses semantics and open dialogue, via API, with multiple Tools rather than acting as a repository of files and models. BIMMS is a new concept of CDE created for the operation/renovation phases in mature real estate markets (such as the European one).

A. Pavan (✉) · V. Caffi · C. Mirarchi
Department ABC—Architecture, Built Environment and Construction Engineering, Politecnico di Milano, Milano, Italy
e-mail: alberto.pavan@polimi.it

A. Valra · D. Madeddu · D. Farina · J. Chiappetti
One Team S.r.l., Milano, Italy

© The Author(s) 2022
B. Daniotti et al. (eds.), *Innovative Tools and Methods Using BIM for an Efficient Renovation in Buildings*, SpringerBriefs in Applied Sciences and Technology, https://doi.org/10.1007/978-3-031-04670-4_3

Keywords Common data environment · BIM management system · Digital
platform · Digital twins · IoT

3.1 Common Data Environment

3.1.1 Introduction to the Topics

The sharing and conservation of information are two critical issues that have always
been present in the construction sector. They have been handed down over the
centuries and have become increasingly important with the increase in regulatory,
technological, requirements, etc. complexity. typical of our days.

Today, digitalisation offers new tools to solve this critical information throughout
the construction chain, both in terms of containers (files) and in terms of content
units (single information and data). With the BS 1192:2007, in the UK, the concept
of the Common Data Environment in constructions is born, then resumed in a more
digital sense in the PAS 1192-2: 2013. Since the CDE has been "subject", of other
standards, at national level (UNI, DIN), European (CEN) and international (ISO),
and "object" of new tools (360DOCS, Aconex, BIM+, etc.). Tools that from simple
document repositories have increasingly turned to the management of the "models"
generated by the various "BIM Authoring" software, in their different component
files (architectural, structural, MEP, etc.).

The writing of a new type of CDE, aimed mainly at the built environment, cannot
therefore ignore a careful analysis of the standards that define its contours and the
commercial software that characterize its operational use.

The creation of a CDE aimed at the construction supply chain, and which sets
the life cycle of buildings (50, 100 and more years), also cannot omit even a general
analysis of the systems for sharing and preserving "data" (and not just files). These
systems have been in use for decades in other industrial sectors (Data Base Management
System: DBMS; Enterprise Resource Planning: ERP, etc.) and must also be
analysed to define the architecture of BIM4EEB's new CDE system. In the same
way one cannot forget all the new technologies that increasingly appear in the informative
panorama also of the constructions: sensors (Internet of Things—IoT), DB
to Objects, non-relational DBs, etc.

Finally, the same concept of CDE addressed to the single order, or intervention, is
now losing more and more meaning towards that of digital platform, collaboration and
sharing: organization, supply chain, national and transnational (see the DigiPLACE
project, EU, H2020).

The architecture of the BIM4EEB CDE defined in BIM4EEB must take all these
variables into account and give them an operational response.

Building Information Modelling (BIM) is gaining momentum in the AEC industry
for design and construction and one of the most significant technological advancements
in recent years that has been adopted by the design and construction industry
(Khaja et al. 2016, Parlikad et al. 2019).

Traditional method is mainly concerned with 2D representation of information throughout all the entire project phase; however, Building Information Modelling leads to the re-shaping of the construction industry as it stands for the necessary evolution of the design approach linked to the increasing complexity of the building process. BIM lays down the transition from unidirectional and asynchronous workflows to integrated and shared models. Most research has agreed that BIM is a process of expanding 3D models to computable nD models to simulate the planning, design, construction, and operation of a facility. In particular, 3D BIM makes it possible to perform specific analysis based on the geometrical information of the model, such as 3D visualization, clash detection and code checking (Solihin et al. 2017).

To put it in a nutshell, the five most important benefits include: better cost estimates and control, better understanding of design, reduce construction cost, better construction planning and monitoring, and improvement of project quality.

Despite many advantages of BIM, the slow adoption to date has inspired researchers worldwide to investigate existing barriers; these barriers include technical problems (compatibility and reliability), fragmentation of the project team, the inherent resistance to change by construction stakeholders, lack of training, and business process related issues, inadequate organizational support and structure to execute BIM, and lack of BIM industry standards.

The Common Data Environment was originally defined in BS 1192:2007, then in PAS 1192-2:2013, finally it has its own international standard, ISO 19650. There are also other standards from Italy and Germany with similar approach. The contents of the CDE are not limited to assets created in a 'BIM environment' and it will therefore include documentation, graphical model, and non-graphical assets. In using a single source of information collaboration between project members should be enhanced, mistakes reduced, and duplication avoided. The advantages of implementing a CDE include:

- Ownership of information remains with the originator, although it is shared and reused, only the originator can change it;
- Project team members can all use the CDE to generate the documents/ views they need using different combinations of the central assets, confident that they are using the latest assets (as are others);
- Shared information reduces the time and cost in producing coordinated information;
- Any number of documents can be generated from different combinations of model files.

BIM teams are those whose possibly geographically dispersed members from various organizations and disciplines, perform project tasks on BIM-enabled projects. even though achieving BIM's full capabilities relies on effective collaboration among the team members in BIM-based construction networks, it is still a struggle for these members to collaborate. Nonetheless, only a few studies have been conducted to identify the barriers to strengthening team collaboration in BIM-based construction networks.

BIM methodology is mainly developed and applied for new building projects. Its use for renovation and retrofitting projects is still in its infancy. On the other side, there is a strong present need to improve the quality and the functionality of the existing building (Scherer 2018). To encourage BIM adoption for renovation project, there is a need to examine refurbishment project stakeholders' separate roles, responsibilities, relationships, and interactions among themselves of which can be hampered by internal or external environmental factors. Although BIM adoption environment requires a more multidisciplinary collaboration effort of different disciplines against information sharing, building design, construction techniques.

3.1.2 Principal of CDE

Traditionally, interdisciplinary collaboration in the fields of Architecture, Engineering, Construction (AEC) is based on the exchange of 2D drawings and documents. Although the different disciplines use 3D models for development, design, and visualization, the collaboration between them has developed especially in 2D. Traditionally used CAD tools have some significant limitations such as the separation between the design and the information associated with it.

The creation of a common portal, a source of information, reduces exchange times and costs because it makes an easy collaboration between team members who can draw from a single source. In the traditional information sharing, each subject exchanges data with all the other subjects involved causing inevitable information losses (Fig. 3.1).

The structured use of a data-sharing environment requires rigorous discipline on the part of all members of the design team, in terms of adherence to agreed approaches and procedures, differently from what happens for more traditional procedures (Fig. 3.2).

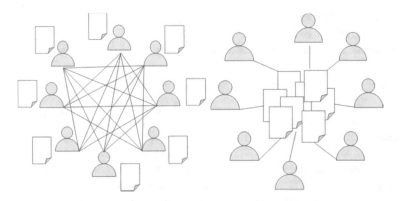

Fig. 3.1 Traditional approach versus ideal BIM-CDE

Fig. 3.2 Stage 3 ISO
19650-1:2017—CDE
schema

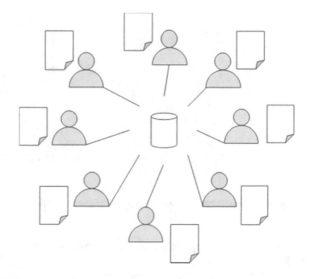

The BIM design was born as an evolution of the process that starts from the design of the 2D paper, to 2D and 3D CAD systems, parametric modelling up to digital modelling in a collaborative 3D environment. The BIM methodology in fact, in addition to requiring the use of specific modelling technologies for the various dimensions, must follow a flow of information that follows well-defined rules and standards.

The main CDE standards are:

- BS 1192-1:2007
- BS PAS 1192-2:2013
- UNI 11337-1:2017
- UNI 11337-4:2017
- UNI 11337-5:2017
- ISO 19650-1:2018
- DIN SPEC 91391-1:2019
- CEN TR 442031/442032
- (ISO 19650-4: WIP).

3.1.3   Practice Applications BIME4EEB–BMS

The DBMSs (Data Base Management Systems) have been created to make a coherence between the independent data and created by different applications. However, this improvement in data management has led to the emergence of independent databases originating from individual organizations that do not talk to each other. The situation remains devoid of a unified vision where data from different databases

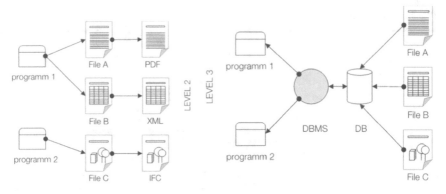

Fig. 3.3 Independent files versus integrated database+DBMS

can communicate quickly and strategically, bringing the process to a higher level (Fig. 3.3).

The ACDat is information content represented by data and data structures (files or containers). Despite wanting to distinguish itself from a file and file sharing system and from a document manager it performs the same function. In general, they are relational databases, but there are ACDat prototypes based on non-relational DBs. The main objective of the ACDat is to guarantee over time the digital sharing of the data created by the various subjects that collaborate in the construction process.

Originally the use was based on the sharing of graphic models (generated by BIM authoring software) within the design team elaborated by the various disciplines: architectural, structural, plant engineering, mechanical, hydraulic, electrical, etc.

"Issues relating to the coordination of (graphic) files, federation of models, verification of geometrical interferences (clash detection), processing of files at the same time by more operators, etc. (see BIM Plus, Trimble Connect, Collaboration for Revit, BIM 360 Team, BIMX, etc.)." (Daniotti et al. 2020).

Today there are platforms that have generated or are evolving into real ACDats /CDEs in which the original function is added to the management of data, files, and documents of various origins; moreover, the coordination will not only be of 3D models but also of time management (4D), costs (5D), maintenance (6D), etc.

Every time a new job is opened, the organization creates a new CDE. The various CDEs (of the various orders) are organised in a collaborative platform. The client is always responsible for the CDE as the owner of the results of the order (Appointing Client, Owner). The designated subjects, who will intervene in the job, interact directly in the CDE. The designated parties and the Lead Appointed parties operate in their own WIP and make their own models and documents available to the other parties. The appointed party approves the documents located in the WIP space and makes them visible in the conditional space for the coordination of all parties. Once the coordination is concluded, the appointed party authorizes the publication (models and documents can be reused by others), and for the appointed party they are ready to be deposited in the Appointing party's CDE.

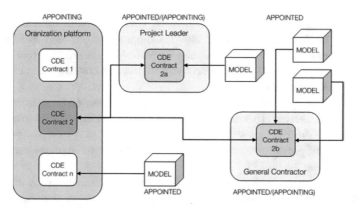

Fig. 3.4 CDE flow

"Therefore, the CDE managed by the Appointing Party receives the definitive data coordinated in advance between the Appointed parties in the Lead Appointed Party's CDE. The entry of the Appointed parties' coordinated data in the CDE job order takes place in the Sharing environment, it assumes value of deposit for the client, and is always subject to the client's approval and verification. The CDE data—coordinated, approved and verified, upon the client's prior authorization—can therefore be shared with third parties (Publishing) or in conclusion archived (Archive) for their following possible mining, should they be re-processed" (Daniotti et al. 2020) (Fig. 3.4).

The main scope of the BIM4EEB BMS is the development, starting from a Common Data Environment, a management platform able to collect and manage the amount of data useful for BIM4EEB project. Such a platform should be open, interoperable and provided of a specific toolkit to optimise the management of information (collection, share, update, and exchange) during the different stages of a BIM-based renovation process. This for making it more efficient with respect to traditional processes. The open and interoperable platform will be based on a BIM management system (composed of different BIM tools) allowing different stakeholders (as designers, architects, construction companies and service companies, owners, and inhabitants) to exchange data and to efficiently manage renovation works.

The BIM4EEB BIM Management System aims to be an open integrated BIM-based collaboration environment, enabling the continuous updating, enhancement, improvement, and enrichment of available models by the AEC industry stakeholders based on robust modelling guidelines that will be provided by the project, to ensure sustainability of the project developments and enhance its exploitation potential even after the end of the project.

The BIM management system will host a variety of applications to ensure timely and uninterrupted realisation of standardised renovation processes and smooth collaboration between the different stakeholders in a cost-efficient and timesaving manner.

Initially, the first one (field survey) will be accelerated thanks to a specific tool for fast mapping that apply the latest information technologies to create BIM models of existing buildings in a quick and accurate way. Then, the second one (initiation and design) will be improved thanks to a tool that support decision-making and a BIM-assisted energy refurbishment assessment tool. Moreover, the third one (construction) will see a BIM-enable tool for the renovation intervention management to organise, plan and track site operations.

The BIM management system will have a set of Application Programming Interface (API) and Services specifically developed to give complete and interoperable access to the BIM4EEB project data. It will be realised as a web platform, and it will serve to all project partners to connect their tools and applications used and developed during the project.

The BIM Management System will be built around a general schema of Common Data Environment (CDE) that will store all the data and information gathered through different sources and along the whole building life cycle, acting a single source of truth (SSOT), with dedicated interfaces and capabilities that will work as a central repository enabling the data exchange with connected tools for providing better coordination among users and process phases.

Common Data Environment (CDE) with dedicated interfaces and capabilities that will work as a central repository enabling the data exchange with connected tools for providing better coordination among users and process phases. All the data will be shared and accessed with pre-established rules, traceable with historical records and revisions, and interoperable through service-based software interfaces. Building related geometry will be stored and viewed as BIM Models using an AEC industry recognized standards as ISO 16739 Industry Foundation Classes IFC.

3.2   Collaborative Workflow and Information Exchange Requirements

The building renovation works, involve several stakeholders, each of them having different and specific needs in terms of information they request and produce for their activities.

To identify how the information workflow has to be organised to support at best the renovation process, it is crucial analyse the relationship among the different actors and their interactions. This becomes even more important when the process is BIM enabled: identifying the information exchange process is a must to fully exploit the potential benefit offered by the new technologies.

This section highlights the relationships and interactions over the whole process, it identifies the actors' profiles and clarifies their interaction. Some general schemes and use cases are presented to highlight how to determine the information exchange requirements and workflow.

The methodology that has been followed to define the stakeholders' profiles and the phases of the workflow they are involved in are described. Visual schemes are presented to highlight sequences and relationships among actors, tasks, phases.

The specific graphical schemes that have been designed to illustrate the relationship at each stage of the workflow, among the actors and with the tool categories are here presented and explained. 21 actors were identified in the former paragraph, using such diagrams. Each profile is accompanied by a table showing the relationship among BIM tools categories that might be involved in the process, the information flow among the actors, the performed actions.

A Use case Matrix is suggested, to establish a general method to define User profiles.

3.2.1 Methodological Approach

To achieve the goals of identifying how the information flow has to be organised to support at best the renovation process, the building renovation process has been analysed through all its stages including initiative, initiation, concept design, preliminary design, developed design, detailed design, construction, use, and end of the life cycle. The steps of the analysis are highlighted in Table 3.1.

The workflow is analysed to define the relevant stakeholders, the relevant input and output information for each possible activity, the related BIM tool available on the market, the priority level of BIM technology adoption for each action or activity. The users' profile for a BIM management system, are eventually determined in relationship to the information workflow, and required function, required BIM tool categories for each stakeholder.

A further study and analysis related to the process workflow has been carried out in order define the relationships between the different actors, how actions are connected and connect the actors and if and how BIM tools are involved in the process, according to a priority scale. The outcome has eventually produced the

Table 3.1 Sequential methodology of the work

Step 1	Collecting detailed data for all activities in all stages of building renovation workflow
Step 2	Identify the stakeholder in each stage, and the relationship between the stakeholders in each activity in the perspective of information analysis. For example, in a single task, who is the information provider, who is the information receiver
Step 3	Identify the input and output information for each action in detail
Step 4	Determine the priority of using BIM technology in each action with the scale from 0 to 5. In detail, 0 is not necessary to use BIM, 1 is useful but not recommended, 2 is recommended, 3 is desirable, 4 is highly desirable, 5 is mandatory. The priority is explained based on research hypothesis and BIM benefit for each action
Step 5	Determine the BIM tools and BIM tool category available in the market that fit for using in each activity in the workflow
Step 6	Based on previous analysis, summary the data for user's profiles including information, BIM category and required function

User profiles identification or a method to identify other User profiles following the workflow, actors involved, required actions, tools categories.

3.3 The BIM Management System

The BIM Management System (BIMMS) is a software platform designed to manage all the buildings' data and to perform as a single source of truth for the stakeholders involved in the renovation project. The platform was developed around a common data environment (CDE) used to store data coming from different sources and along the whole building life cycle. The CDE allow users, to upload, share and manage information containers and provide functionalities to visualize their properties, download, rename and manage document versions. The BIMMS' CDE manages different kind of resources and file formats, and let the user to define properties, set classifications, assign mappings with ontologies according to their role permissions and resource status. A BIM Viewer supports the 3D IFC BIM resources with functionalities to pan, rotate, zoom the models, select geometry items, and view their properties both in 3D and in a hierarchical tree view. The BIMMS' IoT middleware retrieve and save sensor measurements from a Samsung SmartThings Hub (Samsung 2020) collecting the data streaming from the devices placed in the demonstration buildings.

The measurements are stored in the BIMMS and shared, with authorization and permissions defined, to the end-user tools and applications. The BIMMS provides a SPARQL Endpoint to receive and process requests to the resources stored in the CDE using SPARQL queries. The SPARQL Endpoint is compliance with SPARQL 1.1 (W3C 2012) and allow connections to other SPARQL Endpoints to retrieve data from different repositories using the federated queries. To facilitate the visualization of the ontology relationships between classes and properties, the BIMMS implement a graph visualization of the ontologies. Finally, the developers can connect their tools to exchange data with the BIMMS using the BIMMS' REST API.

3.3.1 The BIMMS Architecture

The Common Data Environment (CDE) is the core of the BIMMS, and a service interoperability interface works as a module that enables all the functionalities to be performed like upload, view, manage versioning, sharing, and permissions management. To guarantee the access to the data, an exchange layer service has been developed working as an interface between the external applications to exchange data directly from the application.

The BIMMS database architecture is composed of two main databases, a relational database management system based on MySQL and a hybrid triple store database system based on Virtuoso.

The MySQL database is used to manage and store data for the File management, Users' management, Roles management, 3D BIM Model IFC entities, Building activities, alerts, and occupancy, Sensor's data streaming, Application errors and logs. The Virtuoso database is used to manage and store RDF data for the BIMMS resources, IFC models, sensors data, geospatial data, and the RDF data directly saved by the tools.

The BIMMS provides different functionalities that allow the users to manage all the resources, being unaware about the complex technical workflows happening in the backend.

In a typical workflow the user logs in and is authenticated as profiled account in the BIMMS and then, using the BIMMS' "New resource" functionality, he can upload files, set properties, classification, and mappings with other resources. When a new IFC resource is created, the 3D IFC model is uploaded and the BIMMS, in background, starts a batch process to parse the model and store the items in the MySQL RDBMS, and the RDB-to-RDF conversion. The process does not block the users' activities and can continue to work in the BIMMS. At the end of the conversion, the user will be notified by email about the availability of the BIM Model ready in the resource list.

The resource data is immediately available in the BIMMS' Resource list, ready to be referenced.

When the BIM Model is ready, the user can navigate through the model and view the model's element properties in the BIMMS' BIM Viewer.

The data stored in the MySQL RDBMS are also available in the Virtuoso triple store as ontological representation and stored in specific Virtuoso named graphs.

The data is then available also through the BIMMS' SPARQL endpoint.

3.3.2 Data as Resource

All the contents of the BIMMS' CDE can be considered as resources using the Resource Description Framework (RDF) specifications. A resource is an any physical or virtual entity connected to a computer system. Using the RDF, the resources can be read as a statement in the form of triples (subject-predicate-object). To describe a resource is possible to declare more triples that define the resource parameters, descriptions, and relations with other resources. In the triple, the subject denotes the resource, and the predicate expresses a relationship between the subject and the object. As an example, we can consider as resources different types of entities such as a 3D model, a building element, a user, a task, a sensor, or a document. Each resource will be referenced in the CDE by a unique identifier. This unique reference will be used to retrieve the representation of the resource and to establish links among other resources stored in BIMMS' CDE and across the Web. The RDF representation may contain links to further objects whose content—if considered relevant—can be retrieved by following the links.

The data of the BIMMS' resources, the sensors streaming, and other data stored in the MySQL database is also represented as RDF data through the Virtuoso Linked Data Views functionality (OpenLink 2018). The Virtuoso Linked Data Views is a feature of Virtuoso server that allows to generate RDF views over SQL Data using "machine-readable relations semantics via terms defined in a vocabulary or ontology" (Heward-Mills 2018). The data stored in the MySQL relational database can be available in Virtuoso database as RDF, through a mapping of the MySQL table views with terms and ontologies schemas. This configuration allows to bring all the benefits of using the two platforms and made available as Linked Data the information stored on the relational MySQL database.

The feature can be accessed through an application wizard in the Virtuoso Server that create the mapping definitions in R2RML standard language, that can generate a Quad Map Definition using an ontology based on the data headers available on the source tables and the RDF Schema vocabulary. The R2RML language is a W3C recommendation that define a language that allows to create mappings from relational databases to RDF datasets (W3C 2012). The Quad Map refers to the Virtuoso storage schema that uses quads instead of triples using a field to refer to the source application or resource (OpenLink 2018).

The ontological representation of the resources is defined using the Virtuoso Linked Data View through the R2RML mapping from MySQL logical tables to RDF. The Virtuoso's R2RML processor generate the R2RML default mapping based on the input database schema: the logical tables are mapped as rdfs: Class with the same name, the contents are mapped in data properties defined with owl: DatatypeProperty, and object properties defined with owl: ObjectProperty. The mapping generates an internal ontology that support the generation of the triples composed by a subject map and multiple predicate-object maps, that is used to map each row in the logical tables. The internal ontology is stored in Virtuoso Server and is checked with the OOPS! OntOlogy Pitfall Scanner! (Poveda-Villalón et al. 2014), the University of Manchester OWL Validator (Horridge 2009), and the Vapour Linked Data validator (CTIC 2011). The result is a Direct Mapping that is further customized to include classes and properties defined in the Digital Construction Ontologies (DiCon). The DiCon ontologies are modularised and cover the digitalized construction processes domain. The resources stored in the BIMMS are first covered by the Information ontology module, that provide classes and properties for the information content entities, information containers, designs, plans, events, and issues. The integration was done defining some classes as rdfs:subClassOf and properties as rdfs:subPropertyOf.

For the resources, the mapping uses both classes and properties of the custom ontology based on the MySQL table column headers, the Digital Construction Information ontology that "defines the representation of information content entities in construction and renovation" (Törmä 2020), and the ISO 21597-1:2020 Container ontology (ISO 2020).

For the IFC models, the mapping uses both classes and properties of the ifcOWL ontology (Pauwels 2019), the Building Topology Ontology (BOT) (Linked Building Data Community Group 2021), and the Digital Construction Entities ontology that "defines the basic classes and properties needed for the representation of construction and renovation projects" (Törmä 2020).

For the sensor's data, the mapping uses both classes and properties of the custom ontology based on the MySQL table column headers, and the Digital Construction Ontologies (DiCon) that "act as an enabler of semantic interoperability between systems in the construction and renovation domain" (Törmä 2021).

3.3.3   Working with IFCs

The BIMMS' CDE manages different kind of files that ranges from the 3D BIM models, drawings, reports, spreadsheets, to images and other kind of medias used to store the building documentation. The 3D BIM models are stored using the interoperable IFC file format. The IFC file defines an EXPRESS based entity-relationship model, consisting of several hundred entities organized into an object-based inheritance hierarchy. The structure of the IFC allows to parse the files to extract all the IFC entities, their properties and relationship and save them in the MySQL database maintaining the relationships of the original file.

Although the BIMMS store the original IFC files in the file system, the availability of the content of the IFC files stored in the database improved the speed performance to read and write IFC data, and to retrieve object data in real time to support the visualization in the BIM viewer. Moreover, the applications can define complex queries in the database and retrieve data faster than reading the IFC files on the fly. An application "IFC to Relational DB" runs as a service on the BIMMS' server, that parse asynchronously the IFC files uploaded in the Common Data Environment (CDE) and write the contents in the MySQL database. The application is scheduled and can catch the file version changes, updating the data stored in the database in near real time as the IFC files being updated. The service uses the xBIM Toolkit, a software development toolkit (SDK) that allows to read, write, and view the BIM models in the IFC format (xBIM 2020). Another application "IFC to RDF" runs as a service in the BIMMS' server, to parse the IFC files uploaded in the BIMMS' CDE and to convert them into RDF graphs. The conversion was carried out using an application that convert IFC files in RDF using ifcOWL ontology (Pauwels 2019) and the Linked Building Data (BOT) ontology (Linked Building Data Community Group 2021). The original source code of the application is made by Jyrki Oraskari, Mathias Bonduel, Kris McGlinn, Anna Wagner, Pieter Pauwels, Ville Kukkonen, Simon Steyskaland, and Joel Lehtonen, is also based on some third-party dependencies, and refers to the work of Pieter Pauwels for the main Java components like the IFCtoRDF converter (Oraskari 2020). The source code had been modified to adapt the URI to the BIMMS URI and are automated to be run asynchronously after the IFC has been uploaded in the BIMMS. Once the converted RDF files are available, are stored in the file system, available in the MySQL database as resources, and then loaded in the Virtuoso Server as RDF triples in the named linked data graph.

The BIMMS' BIM Viewer allows to view and navigate through the IFC BIM models, selecting the elements and view its properties. The BIMMS embedded the xBIM Toolkit Libraries to manage the viewing, reading, and modify the IFC files (xBIM 2020). To speed up the visualization, the IFC files are converted using the xBIM libraries in an optimized xBIM webGL format (wexBIM). The webGL format is widely used and supported as web standard to visualize 3D geometry in the web pages (Khronos 2020). As described before, an internal BIMMS' application (namely IFC-to-DB) parse and store the IFC data of the model items in the MySQL database.

The data is used to speed up the visualization of the properties in the BIM Viewer and to support the creation and editing of the IFC spatial elements like IfcZones and IfcGroups. This feature allows to edit the spatial hierarchy of a IFC model adding or modifying IfcZones and IfcGroups used to manage thermal zones for energy simulation, spatial zones for location-based activities, locations for sensors positioning, and element grouping for activities and resource management. All changes are saved in real time in the MySQL database and are available immediately in the BIM Viewer, in the IFC model when downloaded, and in the Virtuoso triple store. All functionalities to add, delete and modify IfcZones and IfcGroups are available via REST APIs for the applications (BIM4EEB 2020). The IFC hierarchy panel in the BIMMS' BIM Viewer with the IfcZones defined to group IfcSpaces for the apartment sensors setup: all IfcZones are defined using the BIMMS' APIs (Fig. 3.5).

Fig. 3.5 The IFC hierarchy panel in the BIMMS' BIM viewer

3.3.4 *Working with Linked Data*

The BIMMS' "Geo Linked Data" functionality allow to define one or more positioning of a resource or a thing in the world through its latitude and longitude. This feature allows to search a resource by its location and retrieve additional information of its neighbour's geographic elements like cities, towns, buildings, hospitals, streets, factories, shops, points of interest, etc. The user can navigate in the map panning and zooming, and then clicking on a specific point on the map can define the positioning. The positioning data is defined with the Digital Construction Entities ontology (dice) developed in WP3, with dice: Location and dice: SpatialPosition and the data properties hasLongitude, hasLatitude, has Altitude, and isLocatedIn. The dice ontology is aligned with the WGS84 Geo Positioning Ontology.

The definition of the location of a resource can be done during the creation of the resource using the BIMMS' "New resource" and "New Linked Data object" functionalities. These functionalities are based on a step-by-step wizard that include a step dedicated to the GeoLinked Data. The GeoLinked Data are then available for consultation through the BIMMS' "Geo Linked Data" functionality that shows all the locations in a map with single points or clusters depending on how much locations are defined near each other. When the point is selected (clicking on it), a new panel will be shown. This panel shows a map with the results of a federated query done using the Wikidata Query Service to retrieve information about geographic elements located near the location. This feature allows to exploit the data distributed on the Web and improve the knowledge of the neighbours where the project resource is located. The query uses the wikibase service in Wikidata Query Service, passing the Latitude and Longitude of the dice: Location as center of interest that, combined with a radius (by default 1 km), will get the results shown in the map as coloured points.

This feature allows to select the points that will open a descriptive label with a few links to the main Wikidata web page where are listed all the property statements available (Fig. 3.6).

The Wikidata Query Service was chosen among others thanks to the high availability of their datasets and for the high reliability of their endpoints and servers. Unfortunately, most other interesting open data sources in the field of geospatial information that were available with their SPARQL Endpoints in the beginning of the project have their services temporarily down due to long maintenance downtime or capacity problems.

3.4 The BIMMS Tool-Kit Integration

The BIMMS REST APIs allow the connection of external tools to the BIMMS environment promoting the development of specific applications that can work in the different areas of the renovation process responding to the specific needs of the

Fig. 3.6 Wikidata query service

stakeholders involved in the process itself. In the BIM4EEB project different tools have been developed according to this schema while the BIMMS REST APIs can be used to develop further applications after the BIM4EEB completion. This approach sees the integration of BIMMS and tools and requires the development of dedicated testing and validation activities that, on the one hand can assess the function of the single tools and, on the other hand, can test and validate the integration between tools and BIMMS guaranteeing the correct information flows between the two.

3.4.1 Tools Definition and Integration with BIMMS

As already mentioned, the BIM4EEB project sees the development of a BIMMS that allows the integration of different tools through REST APIs. The structure of the BIMMS and the integration with the dedicated tools developed in the project can be described through the following image (Fig. 3.7).

The BIM toolkit comprehend six tools integrated around the BIMMS. The tools are following listed.

- BIMPlanner.
- BIMeaser.
- Auteras.
- BIMcpd.
- Fast Mapping toolkit.
- BIM4Occupants.

The first step in the testing and validation activities consists in the identification of the main components in terms of both functionalities of the single tool and integration

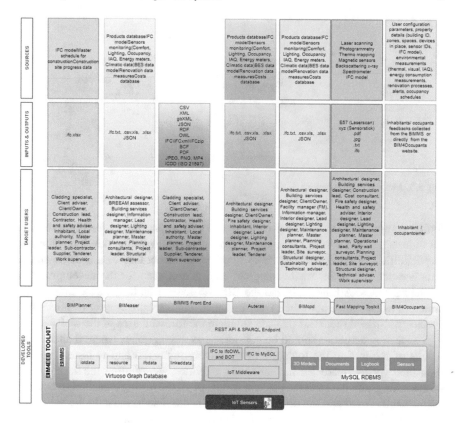

Fig. 3.7 Overview of the BIM4EEB toolkit framework

process between tools and BIMMS with the aims of assuring that these components can fulfil the project KPIs and can work with the expected performance in real environments.

3.4.2 Tools Testing and Validation

The testing and validation activities are based on a structured methodology that contains two main components, namely the testing actions of all the tools, system, and processes of the BIMMS toolkit, and the collection of the feedback by the stakeholders for the different components of the BIMMS toolkit with the aims of identifying mitigation actions applicable for future developments of the system. The main elements of this methodology are depicted in the following image (Fig. 3.8).

According to the series of standards ISO 29119 the testing activity has been organized according to two parallel dimensions. On the one hand performance tests have been developed considering the specific time and resources constrains of each

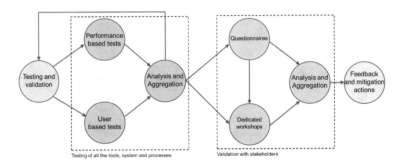

Fig. 3.8 Methodology structure

toolkit component. On the other hand, the so called *"use based tests"* represent a set of tests that are related to the procedure and scenario testing developed through the definition of dedicated test cases documented according to a specific shared template defined in this methodological section and used for the development of all the activities. All the tools and the BIMMS have been tested according to this structure to guarantee both the IT performance and the capability of responding to specific use-based scenarios that can simulate the foreseen uses of the tools.

The testing activities should be considered as circular ones where thanks to the results of the testing the tools are progressively improved and re-tested to reach their final version ready for the validation activities. Hence, after the development of the testing activities and the positive results of all the testing considering both the performance based and the used based ones it is possible to move to the validation activities with the stakeholders described in the following section.

3.4.3 User Feedback and Improvements

The validation activities are based on the development of structured workshops dedicated to the different components of the BIMMS toolkit. The workshops aim at collecting the feedback of the stakeholders that may use the toolkit components to understand their perception in terms of usability, improvement of the performance, collaboration, etc. according to the KPIs identified in the BIM4EEB project. Each workshop has been organized according to the specific functionalities and application of the tools presented in the workshop itself. Nevertheless, to optimize the feedback collection from the stakeholders the workshops will use a standardized questionnaire where the questions are then scaled according to the specific needs and objectives of the workshop itself.

The feedback collection is based on a liker scale to identify the agreement or not of the respondents to specific statements about the presented tools and functionalities (Fig. 3.9).

Fig. 3.9 Linkert scale identification

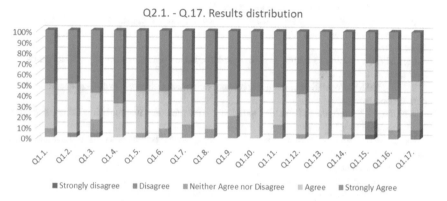

Fig. 3.10 Overview of the questionnaire results

The feedback of the user will be then used to understand if the BIM4EEB toolkit is able to answer to the needs of the involved stakeholders and to identify possible issues that should be corrected improving the toolkit itself. In the following is reported an example of the results obtained from the BIMMS workshop held in November 2021 (Fig. 3.10).

The results clearly represent that the BIMMS responds to the stakeholders' expectations showing an alignment with the majority of the proposed statements. Only four points report a disagreement (even if with a very low percentage—around 3–5%), namely:

- Q.1.2: I find that the User Interface of the BIM4EEB solutions and their user applications have intuitive design.
- Q1.3: Thanks to BIM4EEB solutions I can easily monitor the construction works and schedules during the renovation, compared to a traditional renovation approach.
- Q1.15: I believe that I can use the BIM4EEB solutions with no technical support.
- Q1.17: I believe that my knowledge is sufficient to use the BIM4EEB solutions.

Hance future actions can be activated to provide user guides and resources that can support the stakeholders in using the BIM4EEB toolkit mitigating the highlighted points.

References

BS 1192:2007 Collaborative production of architectural, engineering and construction information. Code of practice

CTIC (2011) VAPOUR a linked data validator. Available at: http://linkeddata.uriburner.com:8000/. Last access 10 Mar 2022

Daniotti B et al (2020) Collaborative working in a BIM environment (BIM platform). In: BIM-based collaborative building process management, pp 71–102

Heward-Mills D (2018) Generating linked data views of SQL relational data with Openlink virtuoso. Available at https://medium.com/virtuoso-blog/rdf-views-generate-b0538101a724. Last access 10 Mar 2022

Horridge M (2009) OWL validator. Available at: http://mowl-power.cs.man.ac.uk:8080/validator/. Last access 10 Mar 2022

ISO 21597-1:2020 Information container for linked document delivery—exchange specification—part 1: container

Khaja M, Seo JD, McArthur JJ (2016) Optimizing BIM metadata manipulation using parametric tools. Procedia Eng 145:259–266

Khronos (2020) WebGL overview. Available at: https://www.khronos.org/webgl/. Last access 10 Mar 2022

Linked Building Data Community Group (2021) Building topology ontology. Available at: https://w3id.org/bot. Last access 10 Mar 2022

Oraskari J, Bonduel M, Pauwels P, Vergauwen M, Klein R (2020) IFCtoLBD. Available at: https://github.com/jyrkioraskari/IFCtoLBD. Last access 10 Mar 2022

Pieter Pauwels WT (2019) ifcOWL ontology (IFC4_ADD2_TC1). Available at: https://standards.buildingsmart.org/IFC/DEV/IFC4/ADD2_TC1/OWL. Last access 10 Mar 2022

Poveda-Villalón M, Gómez-Pérez A, Suárez-Figueroa MC (2014) OOPS! (ontology pitfall scanner!): an on-line tool for ontology evaluation. Int J Seman Web Inf Syst (IJSWIS) 10(2):7–34

Samsung (2020) Smart things developer documentation. Available at: https://smartthings.developer.samsung.com/docs/index.html. Last access 10 Mar 2022

Solihin W, Eastman C, Lee YC, Yang DH (2017) A simplified relation-al database schema for transformation of BIM data into a query-efficient and spatially enabled database. Autom Constr 84:367–383

Törmä S, Zuo Y (2020) Digital construction information. Available at: https://w3id.org/digitalconstruction/0.5/Information. Last access 10 Mar 2020

Törmä S (2021) Digital construction ontologies (DiCon). Available at: https://digitalconstruction.github.io/v/0.5/index.html. Last access 10 Mar 2020

W3C (2012) R2RML: RDB to RDF mapping language, available at: https://www.w3.org/TR/r2rml/. Last access 10 Mar 2022

xBIM (2020) xbim toolkit. Available at: https://docs.xbim.net/. Last access 10 Mar 2022

Chapter 4
Digital Tools for Fast Mapping of Buildings

Cecilia Maria Bolognesi, Eva-Lotta Kurkinen, and Per Andersson

Abstract While the construction sector embraces digitalization, new technologies related to it are spreading benefits. The need of creating a 3D model of a building, a digital copy of something existing, is not new. Mediated by the advent of photographic and laser instrumentation, the construction of a digital model has crossed the fields of surveying with increasing accuracy and precision, imposing standards of capturing the existing first and modelling then ever higher. But while the Building Information Modelling allows a virtual representation of the existing asset enriching its geometry with precious and significant information related to its properties, advanced survey has always faced the impossibility to break the surface of the building, surveying what is inside walls, thus excluding what necessary should be contained within a BIM model. Also, BIM models do not consider the real-time component and do not report the real-time behaviour of the building. In this chapter we will investigate several technologies and instruments exploited till now for the surveying and positioning of existing buildings, plants included, and a new toolkit based on AR that, coupled with sensors and visualisation tools developed by BIM4EEB, offers many advantages when surveying the whole building.

Keywords Advanced survey · Digital · Augmented reality · IFC · Sensors

C. M. Bolognesi (✉)
Department ABC—Architecture, Built Environment and Construction Engineering, Politecnico di Milano, Milano, Italy
e-mail: cecilia.bolognesi@polimi.it

E.-L. Kurkinen
RISE Research Institutes of Sweden AB, Stockholm, Sweden

P. Andersson
CGI Sverige AB, Stockholm, Sweden

B. Daniotti et al. (eds.), *Innovative Tools and Methods Using BIM for an Efficient Renovation in Buildings*, SpringerBriefs in Applied Sciences and Technology, https://doi.org/10.1007/978-3-031-04670-4_4

4.1 Surveying and Mapping the Existing

4.1.1 Existing Praxis, State of the Art, Tools and Workflows for Surveying and Mapping

The existing tools for surveying technologies are multiplying, specialising in relationship to the size of surveying objects and their visibility inside the building also. What can be precisely detected on a surface by one tool can be ignored by another; on the contrary, what is inserted inside a masonry is ignored for example by a laser scan or photogrammetric techniques. Depending on the peculiarities of each one a survey of the building could have to do with:

- Lidar technologies (Light Detection and Ranging or Laser Imaging Detection and Ranging) that allow to reconstruct three-dimensional models through the recording of single or multiple scans determining the distance of an object or surface using a laser pulse, based on the time of flight (the time it takes for the laser beam to travel to the target and reflect back) (Achille et al. 2017); the emitter generates a coded light known to the electronic sensor that strikes the object being measured. The working principle of laser light-based optical 3D measurement sensors can be briefly described as generating a pulse and analysing the signal reflected from the struck object to determine a distance measurement.

  The sensor is defined as active because there is an emitter that emits encoded light, which invests the object to be measured, as opposed to photogrammetry which is passive; in fact, the camera sensor only captures light. Laser emits electromagnetic radiation, it is a wave emitter with time coherence, with same frequency and same phase.

- Photogrammetry, a passive detection system, time consuming in terms of post processing but with undoubted economic advantages; now post processing times and increasingly faster considering the development of the latest modelling software algorithms (Grilli et al. 2017).

- Infrared thermometers as well as heat cameras where thermal mappings give different information about the building such as objects behind others with uniform temperature, find heat leaks, or detect faulty electric cabling and which can rarely be associated with model geometries except with the intervention of the operator (González et al. 2020).

- Radar which uses electromagnetic radiation in the microwave band (UHF/VHF frequencies) of the radio spectrum and detects the reflected signals from subsurface structures.

- Ultrasounds where ultrasonic range sensors produce a beam of ultrasound that is sent out and reflects from the object, allowing the sensor to measure distances but different density of a wall as well. This technology is the same used in medicine application to create multidimensional images of the human body, detecting different densities.

- Magnetic or x-ray sensors but also capacitors or voltage or stud detectors as well as x-rays.

Considering the different structure and nature of data acquired it is natural to investigate a digital database as the structure that can possibly collect them; in the AEC sector a geometrical model is the unique container and the digital replica of what we can investigate. Therefore, difficulties derived from the measurements with different tools are not the only issue; the need to match and position the measurement we are performing is the other one.

Spatial positioning of the model within a Coordinate System and an Internal Positioning System of the surveys carried out becomes a challenge.

In the case of heat cameras matching images in the geometric model can be done manually. In the case of some laser scanners the connected applications allow the automatic recording of the user's movements from one scan to another for pre-recording in the field without manual intervention but the acquisition of associated images in HDR (High Dynamic Range) becomes an automatic process. However, in the case of more precise detection positioning, it may be necessary to scale down the indoor positioning with tolerances of just a few mm.

Almost all indoor positioning systems also lack external reference systems due to their nature and are anchored to temporary positioning systems that should be linked to more general systems.

The GPS geolocation system is particularly effective in open spaces but inside buildings or in heavily urbanised areas, the GPS easily loses operation, and it is necessary to play hard to find other alternatives.

Among the most used technologies for indoor positioning, we consider with different functionalities:

- Beacon (using Bluetooth Low Energy) as other technologies or app for mobile devices have found wide use both in maintenance and in areas more related to Cultural Heritage or Marketing Proximity as tracking tools. Specifically, Beacons are hardware devices thanks to which Bluetooth technology is used to send and receive signals within short distances. They are used nearby as access-points to calculate where the device is located (Pavan et al., 2020).
- Glasses: HoloLens (Hubner 2020), HTC Vive, Oculus Rift. They are different in functionality with respect to which they have been designed but all three disposals contain positioning capabilities with respect to indoor environments. HoloLens, designed for AR for the vision of holograms superimposed on environments, includes several sensors, to measure inertia also, a light sensor and four cameras for environmental analysis. The tool measures the time of flight for IR light and creates a 3d image of the room. The accuracy is about 3 cm. HTC Vive uses an IR-flash, little microchips with a photocell that are tuned to listen to infrared light, followed by sweeping IR-line lasers horizontal and vertical. It is a passive sensor sensing on headset, handles and trackers. Positioning is then determined by detecting the pulses of a photo led. The time between the initial pulse and the pulse generated by the line laser helps the positioning with a 2 mm accuracy with

a 30 Hz update time. There are some limitations in the dimension of the spaces that must be related to the sensor's possibility of detection. Oculus rift uses LED markers on the headset, handles and trackers and then uses cameras to track their position. The cameras are both fitted in the headset as well as the on stands in the room. The accuracy of this system is around 3.5–12 mm but just as with the HTC vive errors increase on distance to the cameras (Weinmann 2021).

- QR codes attached to objects is the most traditional tool we can use but still one of the most accurate. A simple camera tracks objects when the QR code is attached to a fixed object as a reference point. Doing so that position will be geographically known for a camera. The orientation, angle and the position of the camera can be calculated by registering the symmetry and size of the QR-code in the image. Image recognition is used in AR-tools, allowing to anchor/position a virtual model to the real world so that they match each other.

4.1.2　The Survey Process in BIM4EEB

The whole process of BIM4EEB (Daniotti 2021), is structured around the idea of a BIM model enriched with data from different areas of data collection; the data base collects both data coming from users than related to energy needs and consumption and is located in an IFC file in the BIM Management System. However, the model used, a digital twin of the existing, must be reproduced with extreme accuracy to be able to intervene in a punctual way for the renovation process.

One of the tools developed by the project relates to the ability to systematise a variety of existing survey tools to produce a more complete and rapid mapping of buildings. The coupling of different instruments for different purposes has allowed us to realise a new tool used to deepen and speed up the survey.

Installations within walls can be easily detected by the new tool and reconstructed with augmented reality within the realised IFC model.

At this point, the following definitions should be given:

Virtual Reality (VR): VR is an immersive experience based on realistic 3D contents, sounds and other sensations to replicate a real environment or create an imaginary world that you can view through glasses.

Augmented Reality (AR): AR is a live view of a real-world environment with augmented and superimposed contents. The augmentation is achieved utilising devices like smartphones, tablets or custom headsets and dedicated apps that overlay digital contents onto the scene real environment (without interaction).

Mixed Reality (MR): MR—or hybrid reality, is the merging of real and virtual worlds to produce a new environment where physical and digital objects interact and cooperate in real-time.

In our experiment the AR overlapped to the digital model, i.e. a point cloud, is modeled and geolocated starting from the survey that the sensors have captured.

The creation of an AR-tool for fast mapping has been divided into development of a sensor stick with several functions for detecting installations inside walls to be coupled with IFC-file coming from general point clouds previously surveyed.

AR visualisation is used to give detailed information about a placed virtual object, in this case hidden pipes and cords inside the walls and materials. The data from the mapping is shaped as an IFC-file and is transferred to the BIM Management System to be used in a larger workflow for renovation processes.

The AR-tool has been developed to collect sensor data, laser scanning data and mix the data in a 3D-model environment in Hololens.

AR-tool development has been an integration work, divided into development work of sensor-stick, testing of measurement results from sensors in different environments, evaluation of testing data, decision making of methods and choice for type of sensors. The results from the work have confirmed it can be possible to show objects inside walls as electricity cables, humidity, metals, studs, know and locate the position in the building.

4.2   Functionalities of a Fast-Mapping Toolkit

4.2.1   The Hardware

The end goal of the fast-mapping toolkit is to produce an IFC-file that contains information about the building's construction, geometry, and installation. For years now, there has been a consolidated awareness that an accurate and complete 3D digitization is indispensable for various maintenance activities. In our project the IFC-file collects information from a point cloud, a sensor stick used through HoloLens 2 device (AR glasses) and a laptop (Fig. 4.1).

Using surveyed 3D data in the field can facilitate the interpretation and fruition of geometrical shapes in real-time; the opportunity to consult surveying results overlaid to real scenarios improves the building management process. Maintenance actions

Fig. 4.1 The IFC-files are created out of information from a point cloud, HoloLens 2 de-vise (AR glasses), a sensor stick and a laptop

could also benefit from VR/AR/MR solutions and the main idea is to apply concepts taken from Industry 4.0, where AR and MR systems are used to reduce production costs, increase efficiency, and ease working processes faster.

In the fast-mapping workflow the point cloud obtained with a TLS campaign will be visualised in HoloLens two, where Augmented elements surveyed with sensor stick are created. HoloLens are a powerful device, with some computational and battery life limits. They can handle and render 600,000 points inside its field of view before a noticeable frame drop. Battery life is set around 2–3 h, depending on the workload. For this reason, it would be impossible to use, visualise and interact with full-resolution 3D datasets composed of hundreds of millions or even billions of points—typical of large surveys, and this is a limit. The HoloLens 2 can recognize hand gestures, gaze and voice commands. When a user's hand approaches the field of view of the HoloLens 2, one of its depth sensors begins to monitor the user's hand.

Any laser scan can be used to create a setup that is then aligned to create the point cloud; BIM4EEB project tested three demo cases in three different residential buildings, verifying the usability of the system at the scale of an apartment. All the difficulties that can be found in a normal survey remain: the presence of visual impediments to the survey of the normal geometry of the walls, the limited size of some rooms, the presence of reflecting surfaces.

When the point cloud is done, the scan is then imported to the toolkit using a laptop running the so-called Companion App. The Companion App has been developed to prepare the point cloud to stream and to make it available to the point cloud streaming service that is also running on the laptop.

All the workflows need to be accompanied by a wi-fi connection and this may be considered another limit.

The laptop is connected to Wi-Fi to allow other devices to stream and view the point cloud in the same environment.

The point cloud is visualised inside the HoloLens and becomes a reference for creating accurate geometries manually. In order to do that the user on the HoloLens basically creates a specification of the geometry that is sent to the laptop. Another service on the laptop takes this specification and translates it to an IFC-file. The results are sent to the HoloLens device and visualised. Together with all this we have the sensor stick that is also connected to the same Wi-Fi, the HoloLens connects to the sensor stick pulling data from it, translated to the correct position relative to where the HoloLens sees the sensor stick, that creates a position sensor cloud (Fig. 4.2).

4.2.2   The Companion App

The Companion app is written in the programming language Unity and has the task of communicating with the HoloLens and sensor stick. To be able to handle big amounts of data in a Hololens2 device with limited computer resources the calculation parts of the program are put into the Companion app instead of the Hololens2 glasses. This has created a big stream of data between Hololens2 to Companion apps and it

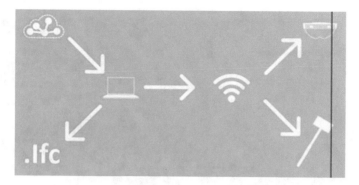

Fig. 4.2 The companion app prepares the point cloud to streaming and makes it available to the point cloud streaming service that is also running on the laptop. The laptop is then connected to a Wi-Fi that makes possible for other devices to stream and view

was decided to use a standalone WIFI. As already said HoloLens2 are small devices and fail in handling big point clouds; to handle big point clouds in a small device like this was created an algorithm that worked together with HoloLens2 device. The algorithm shows the position of the HoloLens2 and its direction part view into the Companion App.

The algorithm then streamed that part of the point cloud that the HoloLens2 devices just needed. By this algorithm we could use a big high density point cloud of a big building into a small device with limited resources. The Companion app has three tabs in the menu; Settings, Ifc Files and Point-Clouds.

Under settings, you set the IP addresses and ports for the HoloLens and sensor stick. You also name where (in which map) point cloud will be picked up and where IFC-files will be stored.

The point cloud and IFC-files are visualised in the Companion App. It is also possible to see where the HoloLens are. The below figure shows the Companion App when the point clouds at CGIs office in Sweden are visible. The red glasses in the left lower corner shows where the HoloLens are and in which direction it is looking (Fig. 4.3).

4.2.3 The HoloLens 2 Device

The training with the Point Cloud started with Hololens1 but shifted to HoloLens 2 as soon as we realised the new more powerful device with the new hardware had more power and features. The new interaction model was also something that was greatly appreciated, the HoloLens2 can track both hands fully which then means that we can have richer, more life-like experience. Interactions with the menu will be more natural and the need for a secondary hand controller we earlier use is no more. But as the HoloLens2 was introduced, some problems did arise, some expected and some

Fig. 4.3 The Companion app when the point cloud at CGIs office in Sweden is visible. The red glasses show where the HoloLens is

unexpected. To be able to use them in the fast mapping they need the App "PolyTech" which is written in Unity and makes it possible to download the point cloud, run the sensor stick and create IFC-files. The menu in the PolyTech app contains; settings, point cloud, IFC-files, the sensor stick, workspace (where the IFC-objects are created) and home (Fig. 4.4).

When creating an object in the HoloLens, it's possible to choose between; 4 Point cube, 3 Point cube, Simple Cubic and Spline. Once objects are created you can

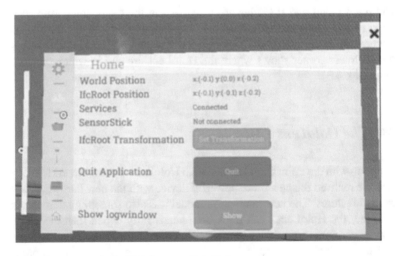

Fig. 4.4 The home menu in the HoloLens app PolyTech

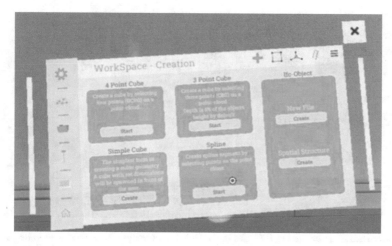

Fig. 4.5 In the workspace it's possible to create objects. You can choose between; 4 point cube, 3 point cube, simple cube and spline

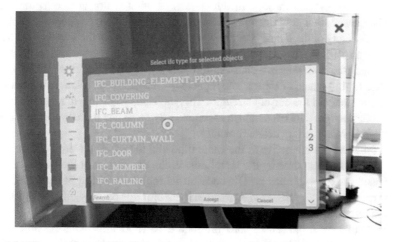

Fig. 4.6 When creating an IFC-object, you need to choose which kind of object types it should be

change position, rotation, and size of it. When creating the objects, you also select which type it shall be. There are different options to choose between. In the figure below IFC_BEAM is selected (Figs. 4.5, 4.6 and 4.7).

4.2.4   The Sensor Stick

The sensor stick is the tool used for the survey of hidden data; it is part of the fast-mapping toolkit together with a laser scan and the HoloLens. It is a hardware

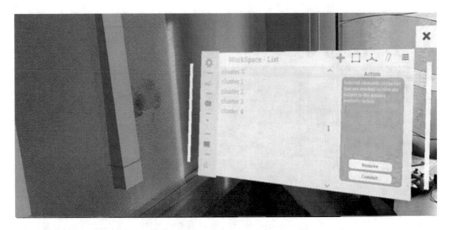

Fig. 4.7 Beams are created as IFC-files in the HoloLens (orange). Now they need to be tilted in the correct position so that they are behind the wall

instrument that detects through different sensors what is not visible to the naked eye. The sensor stick contains four different sensors to catch: temperature, voltage, inductance, and conductance. The temperature sensor detects the temperature at the surface of an object, of a wall, floor, or ceiling. The voltage detects electrical AC 50 Hz voltage at different depths in a construction. It is possible to change the sensitivity of the metre to find what is desired. The inductance measures the relationship between a magnetic flux and the current strength. This indicates where there might be beams in the construction. The capacitance measures the resistance in the material and can in that way among others detect moisture.

All four sensors are active at the same time and the measures are both recorded and visualised in real time by the HoloLens. When the camera at the HoloLens sees the sensor stick (which has an QR-code on its top) its position is registered. The result from the sensor stick is visualised in the HoloLens both by different collars at the point cloud and in the menu as diagrams.

The results from the sensor stick are then used to identify where beams, water pipes etc. are located. The IFC objects are then created with help of the sensor sticks results within the HoloLens. The figure below shows the result from the sensor stick as blue and orange colours. With the colour indication, the beams are then created (Fig. 4.8).

4.2.5 Modelling Objects in a 3D-Environment by a Unity Based AR Visualisation

All interactions with users using a HoloLens2 is based on the users' hand moves that the camera in the HoloLens captures. After scanning with a laser scanner and

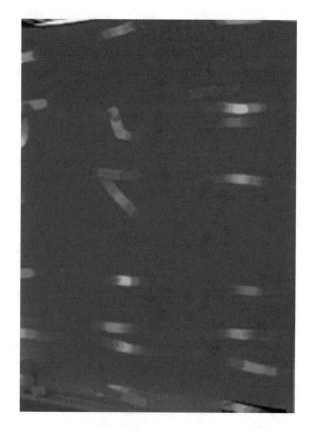

Fig. 4.8 The results from the sensor stick are visualised at the point cloud as different colours. Each sensor gives a specific colour. Here the green colours indicate where the beams are in the construction

the sensor stick you can create the 3D Ifc image of the single room and, adding one to the other of the flat and of a building. The HoloLens could then put out an Ifc building element based on the point cloud and the data from the sensor stick scan. The user is then able to choose which Ifc element to use and then put it into the area based on that kriterium. For example, the application based on the data inside the wall could automatically put out the beams inside a wall. All elements can be moved in all directions and rotated inside the application from the device.

4.2.6 Main and Open Issues

The software is working well but there seems to be some things that could make it more efficient to use after some more development. The workflow has shown that interaction between different data is possible through the interfaces we create with Ifc data even if it comes from different sources, whether laser scan or sensor stick. However, the manual recognition of different objects for the reconstruction of elements in Augmented reality in the long run can be complex and cumbersome.

A possible application to further speed up the survey process could be to develop machine learning processes that would allow us to automatically recognize the most recurring parts within a building and be able to create templates of doors, windows, openings that are placed often, so we don't have to redo as much of the mapping. Also, more improvements should be directed to the positioning of the augmented elements in the walls as it seems that is not as accurate as it should be in terms of positioning.

References

Achille C et al (2017) Learning geomatics for restoration: ICOMOS summer school in Ossola Valley. Int Arch Photogramm Remote Sens Spatial Inf Sci XLII-5/W1:631–637. https://doi.org/10.5194/isprs-archives-XLII-5-W1-631

Daniotti B, Bolognesi CM, Lupica Spagnolo S, Pavan A, Signorini M et al (2021) An Interoperable BIM-Based Toolkit for Efficient Renovation in Buildings. Buildings 11(7):271 https://doi.org/10.3390/buildings11070271

Gonzales et al (2020) An unsupervised labeling approach. Int Arch Photogramm Remote Sens Spatial Inf Sci XLIII-B3–2020:407–415. https://doi.org/10.5194/isprs-archives-XLIII-B3-2020-407-2020

Grilli E et al (2017) A review of point cloud segmentation algorithms. Int Arch Photogramm Remote Sens Spatial Inf Sci. XLII-2/W3:339–344. https://doi.org/10.5194/isprs-archives-XLII-2-W3-339-2017

Hübner K, Clintworth Q, Liu M, Weinmann S (2020) Evaluation of HoloLens Tracking and Depth Sensing for Indoor Mapping Applications. Sensors 20(4):1021 https://doi.org/10.3390/s20041021

Pavan A et al (2020) BIM Digital Platform for First Aid: Firefighters, Police, Red Cross. In: Daniotti B, Gianinetto M, Della Torre S (eds) Digital Transformation of the Design, Construction and Management Processes of the Built Environment. Research for Development. Springer, Cham. https://doi.org/10.1007/978-3-030-33570-0_25

Weinmann S, Wursthorn M, Weinmann P, Hübner (2021) Efficient 3D Mapping and Modelling of Indoor Scenes with the Microsoft HoloLens: A Survey. PFG – J Photogrammetry Remote Sens and Geoinf Sci 89(4):319–333 https://doi.org/10.1007/s41064-021-00163-y

Chapter 5
Digital Tools for HVAC-Design, Operation and Efficiency Management

Teemu Vesanen, Jari Shemeikka, Kostas Tsatsakis, Brian O'Regan, Andriy Hryshchenko, Eoin O'Leidhin, and Dominic O'Sullivan

Abstract The project BIM4EEB aims also to develop digital tools to support the design, procurement, installation, post-renovation operation, user feedback and profiling of building automation systems for HVAC. This helps supporting decision making, interaction with tenants and owners during the design, construction, and post-renovation operation phases. The development of the tools will be underpinned by a sound methodological approach. Work will include considerations of interoperability with Smart City technology of automation systems for HVAC. Specific objectives will be related to the development of the following software tools:

- A software component supporting the automatic generation of the layout for control systems emphasising on user preferences and including constraint checking of BAC-topologies against selected building codes. Data and information stored in BIM models are used to generate the initial recommendations and constraints and to deliver the final installation instructions.
- A software component allowing the seamless specification and evaluation of user comfort and systems performance. The underpinning information model will merge data sources from BIM (dimensional data) and BAC (factual data).
- An energy-refurbishment assessment tool, for bridging the gap between commercial simulators and the BIM management system.
- A user-profiling component allowing to compare expectations of tenants and owners regarding comfort and systems' performance against monitored parameters. The results of this software component can be used in the pre- and post-renovation phases to update the content of BIM systems and thus to improve their accuracy and to reduce efforts for data acquisition and verification.

T. Vesanen (✉) · J. Shemeikka
VTT Technical Research Centre of Finland Ltd., Espoo, Finland
e-mail: teemu.vesanen@vtt.fi

K. Tsatsakis
SUITE5 Data Intelligence Solutions Limited, Limassol, Cyprus

B. O'Regan · A. Hryshchenko · E. O'Leidhin · D. O'Sullivan
University College Cork (UCC), Cork, Ireland

© The Author(s) 2022
B. Daniotti et al. (eds.), *Innovative Tools and Methods Using BIM for an Efficient Renovation in Buildings*, SpringerBriefs in Applied Sciences and Technology, https://doi.org/10.1007/978-3-031-04670-4_5

Keywords HVAC · Decision making · User feedback · BAC

5.1 Constrains Checking, Performance Evaluation and Data Management in Building Renovation Processes

In any renovation project time is an important factor, the ability to have an early indication of a building's requirements, in terms of quantities and positions, of HVAC, lighting and other equipment, can significantly reduce costs and time. In addition, performance evaluation and efficient management of data is essential to the monitoring of any project. As part of BIM4EEB, the BIMcpd Platform was developed, and it contained features to help address this.

The constraint checking module specifically focused on providing an overview of the locations of diffusers, ducting, lamps, sockets, and electrical conduits. Users were able to upload 2D images or 3D models, create zones and select products from a catalogue database. The recommended positions for diffusers, lamps and sockets were calculated, with Dijkstra's algorithm being applied after this to find the shortest path to connect these back to a central point. This offers huge potential to combine with other applications such as computational fluid dynamics (CFD), VR/AR and quantity surveying, to carry out further analysis on the recommendations (Fig. 5.1).

To ensure that the renovation process delivers the savings (energy, carbon emissions and cost reduction) data needs to be gathered, stored, processed, and analysed. In BIMcpd, two modules were used for this: (a) data management and (b) performance evaluation. Data management consisted of two options for the users, file upload or the API which was connected to the BIM Management System

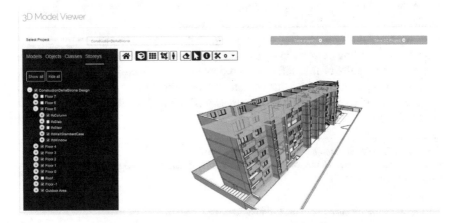

Fig. 5.1 Constraint checking module: model viewer

(BIMMS). Once uploaded/connected, the user mapped the data to the BIMcpd structured database, which ensured that the data being uploaded to the database included essential metadata (such as unit type for temperature). The structed database is core to the performance evaluation module and guaranteed that the information displayed in this tool was accurate (Fig. 5.2).

The performance evaluation tool consisted of several features and was designed with the end-user in mind. Firstly, the user did not have to choose which way to display the data, BIMcpd used an algorithmic approach to determine which way was best. Users could query the database without any experience of database querying languages by simply choosing from a series of drop-down lists. Once displayed, they could apply one of many outlier detection algorithms included in the tool to identify, and if they choose, remove them, and reload the visualisation (Fig. 5.3).

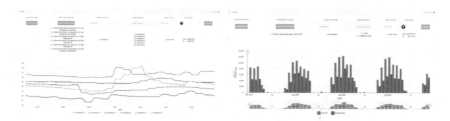

Fig. 5.2 Performance evaluation module

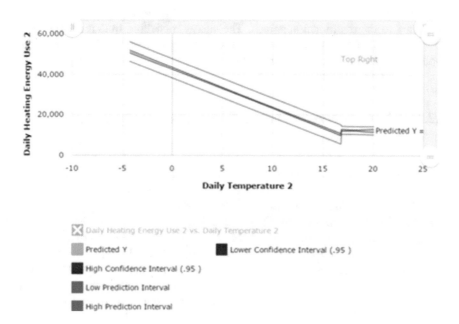

Fig. 5.3 M&V change point regression

Measurement and verification (M&V) is the industry standard for energy efficiency projects, BIMcpd contained a full-featured M&V tool, which enabled users to create a baseline (simple linear, change-point or multi-variate regression) and then once the renovation works have been completed (i.e. application of the energy conservation measures or ECMs), re-evaluate by creating a reporting period and comparing this with the baseline to determine if the expected reductions/savings were achieved.

In any M&V project, non-routine adjustments can be made, due to changes in circumstances between the baseline and reporting periods, BIMcpd contains this option also. Therefore, BIMcpd can help reduce time, costs, and maximise data use in renovation projects.

5.2   BIM-Assisted Energy Refurbishment Assessment Tool

The BIMeaser (BIM Early-Stage Energy Scenario) tool is a tool that supports the energy related decision-making process in the early design stage of the renovation process. The tool enables the assessment of several energy refurbishment design options enabling architects and engineers to provide solutions that best fit to the client requirements while optimizing the energy use and comfortable indoor climate conditions for the occupants. The intended use event of the BIMeaser is the collaborative design session, where building designers (architect, structural, HVAC, electricity) discuss about the expected energy saving measures of the renovated building.

One of the important benefits of the BIMeaser tool is the faster energy modelling compared to the traditional approach. The energy modelling time reduction is − 75% compared to the traditional manual description of the renovated building in the sophisticated energy simulation program.

5.2.1   The Collaboration in the Energy and Indoor Climate Design Process

BIM4EEB BIMeaser tool enables easy build-up of the "As-is" energy and indoor climate model of the building, applying the renovation scenarios and presenting the impact of each renovation scenario. The targeted user role is "Energy expert", who is a separate consultant or a member of the design team. The simulated renovation scenario results will enrich the BIM Management (BIMMS) system content. These will be stored in RDF compatible format into the BIM Management system containing links to IFC model used in the simulation. The linking of OPR's and the BIM model in the BIM Management system enables tracking of the building energy performance during the evolution of the renovation design, which is an important part of the performance-based design approach.

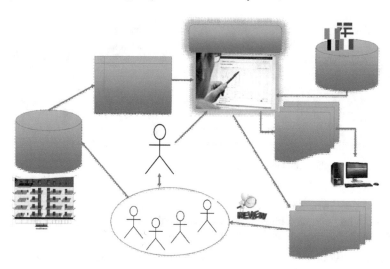

Fig. 5.4 Overview of BIMeaser data exchange

The OPR's—e.g., operational energy cost, payback time of renovation and summer thermal comfort—are an essential part of the performance-based building design process, which assumes that design selections are validated against the OPR's in each design stage before moving to a following design stage. The design team will handle the detailed technical energy selections affecting to the OPR's using the tool as part of the collaborative work (Fig. 5.4).

5.2.2   Tool Operation, Outcomes, and Benefits

The BIMeaser tool and the supporting national renovation measure database was implemented as a web application. The commercial IDA Indoor Climate and Energy 4.8 simulator was connected as a back-end solver with the help of the available API. The technology readiness level of the tool and the API-integration is TRL6.

The intended use event of the BIMeaser is the collaborative design session, where building designers (architect, structural, HVAC, electricity) discuss about the expected energy saving measures of the renovated building. The BIMeaser is connected to the BIMMS, which contain the digital model of the building to be renovated. The digital model is imported to the BIMeaser from the BIMMS. The designers agree several renovation options, which are modelled as energy scenarios by using the supporting "drag and drop" functionality in the tool. The tool contains a national database for the "ready-made" renovation measures, which support the easy application and remembers the previously added measures for further use. Also new measures can be defined into the database if none of the existing ones are suitable. The simulation can be started and after the finalisation of the energy simulations, the

Results

OPR's	Cost			Energy					Comfort	
Scenario	Operational energy cost €/floor-m²ₐ	Payback time Years	Investment €/floor-m²	Primary Energy kWhpr/m²ₐ	RES share %	Heating kWh/m²ₐ	Cooling kWh/m²ₐ	Electricity kWh/m²ₐ	Summer thermal h/year,zone (Tindoor > 27 °C)	Summer thermal °h (Tindoor > 27 °C)
Baseline Via Birona Monza	14.87		0.00	198	0.01	123	0.0	35	1763	4878
Scenario Insulation–211a3c72-67df-4f5d-ae45-11c3ae8e6c73	13.39	31.4	46.41	171	0.01	97	0.0	35	2045	6108
Scenario Solar PV panels–4aee83d0-5df5-4edb-a3ee-1a93b0a75477	12.89	5.7	11.25	198	5.55	123	0.0	26	1763	4863
Scenario New condensing gas boiler–92acc208-2dce-4458-b9b5-8e59bfe79570	13.89	16.2	16.00	180	0.01	106	0.0	35	1763	4878
Scenario Replacing the windows SOSTITUZIONE SERRAMENTI ESTERNI–6438aeaf-578b-4d3f-9934-68705cdadbe9	13.60	56.8	72.18	174	0.01	101	0.0	35	1875	5429
Scenario–7634d9d3-0a94-4c7a-a2cd-f00ea2890a70	9.55	27.4	145.84	136	8.89	64	0.0	26	2284	7263

Ok

Fig. 5.5 An example of the OPR result summary in the BIMeaser tool

values of the Owners Project Requirements (OPR's) can be reviewed. Finally, the OPR's can be published back to BIMMS for further use of other services (Fig. 5.5). The main functionalities of the BIMeaser tool are:

1. Easy build-up of the "As-is" energy and indoor climate model of the building by using the BIM and linked data for accurate modelling in the early design stage (Concept design & Preliminary design), where the most important design selections are made according to the costs and performance.
2. Apply the renovation scenarios to the "As-is" -building. The BIMeaser tool enhances the collaborative work of the design team in the early stage of the design, which usually lacks the sophisticated indoor climate modelling tools. The indoor climate and energy design is a multi-domain challenge, and it should always be considered as a teamwork.
3. Present the impact of each renovation scenario in terms of Owners Project Requirements (OPR).

5.3　Energy-Related Behaviour Profiling Mechanism

One of the main innovations related to building management process is the incorporation of building occupants in the overall analysis. Building occupants are the active entities on the building environment and thus a thorough analysis of this interaction should be considered in any building related decision making. In this section, we present an innovative data driven framework towards the extraction of user-centred knowledge about the building environment. More specifically, the objective is to establish an occupant's context-aware behaviour modelling engine, which will continuously monitor and learn transparently the operational and behavioural patterns (i.e., the preferences) of building occupants, while on the same time interact with the occupants in an ambient manner to define user preferences and extract comfort levels, while taking into consideration also health boundaries. In addition, the overall framework will enable the delivery of Context-Aware Energy Behaviour Profiles, reflecting occupants' energy behaviour as a function of multiple parameters, such as time, environmental context/conditions occupant comfort preferences and health/ hygienic constraints etc.

5.3.1   *Occupancy and Behavioural Profiling Modelling Framework*

The occupancy profiling modelling framework is targeting on extracting near real-time information about occupancy presence by processing data coming from different types of sensors in the building. On the other hand, comfort profiling is defined as a non-parametric model which consider the contextual conditions and user preferences to extract dynamically updated information about users' comfort preferences. In more details, the details of the algorithmic framework towards the extraction of Occupancy and Behavioural Profiling are provided in the following.

Occupancy diversity profiles: The scope of this module is to enable the extraction of accurate occupancy profiles. To exploit the full set of data available, a twostep approach is adopted. At first, initial profiles are defined by an expert and consist of the configuration occupancy file for a building zone. Information about typical occupancy profiles is available at software libraries of professional BIM software. Then, user defined information is incorporated in the analysis. First, the end users of the building may be provided with a tool (user application) to fine tune the accuracy of profiling information. Then and if sensor data available, occupancy diversity profiles are extracted by incorporating actual measurements in the analysis. The overall workflow for the extraction of occupancy profiles is presented in the following figure (Fig. 5.6).

Comfort analytics engine: The process includes data acquisition of sensorial data from different end points. More specifically, indoor temperature, humidity etc.... data are tracked from sensorial equipment installed in building environment (Fig. 5.7). Outdoor environmental conditions can be available either via outdoor sensors installed or via weather service available in public.

In addition to contextual conditions, the comfort analytics engine handle information (control actions) available from the different controllable devices in building premises (i.e., HVAC) as a means of user's settings associated to its comfort mode. This requires special equipment to be installed in building premises. If no special IoT equipment is available, the building occupants can express their comfort settings against the environmental conditions through an intuitive user interface (UI)

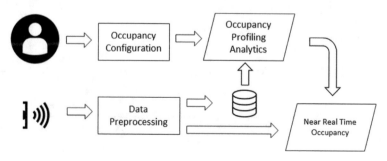

Fig. 5.6 Occupancy analytics process

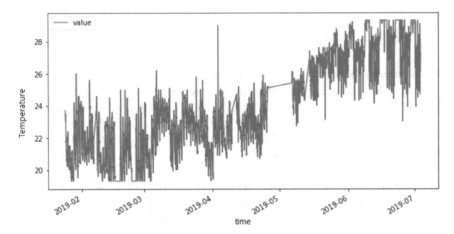

Fig. 5.7 Indoor environmental data—temperature timeseries

developed in the project. A Likert scale approach is considered to get end users' feedback.

Then by applying ML based techniques (MultinomialNB, GaussianNB or regression techniques) we can extract the level of preference or non-preference of an occupant or group under specific contextual conditions. Among the different classifiers examined, the Random Forest classifier is performing the best results (70.7% accuracy level) for the definition of comfort/discomfort values. The results of the statistical analysis over the data reveals the typical thermal comfort profile for the user of the zone, as a probabilistic function of being at a thermal comfort state under different environmental conditions (Fig. 5.8).

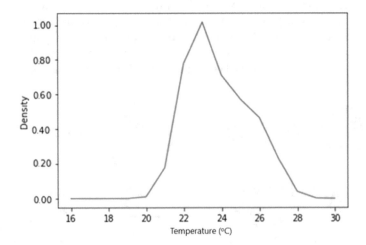

Fig. 5.8 Thermal comfort profiling analysis—distribution curve

5.3.2   Energy Behavioural Profiling Framework

The occupant's behavioural profiles should be further complemented by energy
behaviour profiles with the aim to deliver occupants energy behaviour profiles,
reflecting multiple parameters such as time, environment conditions, energy costs,
occupant comfort preferences etc.... More specifically, indoor temperature and
humidity data, energy consumption, sub metering consumption are tracked from
sensorial equipment installed in building environment (Fig. 5.9). Outdoor environmental conditions can be also available either via outdoor sensors installed or via
weather service available in public.

Then, different ML based techniques (MLP neural net regression Least Absolute
Shrinkage Selector Operator linear regression) are considered to correlate energy data
with environmental and operational characteristics towards the extract of information
about typical energy profiles. This information may be further utilized for the extraction of accurate energy forecasts or energy simulations at the building environment.
More specifically, the following correlations are derived from the analysis:

Correlation versus time. Typical month, week and daily profiles are derived from the
analysis showing the periodicity of the energy consumption over the time (Fig. 5.10).

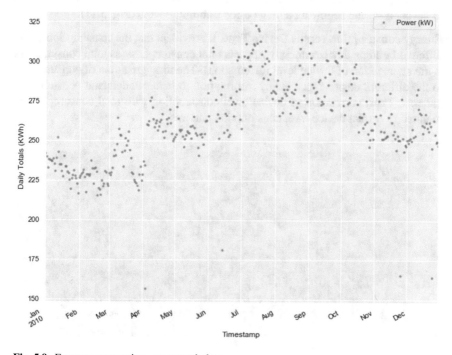

Fig. 5.9   Energy consumption—per month data

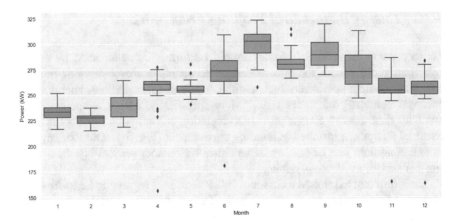

Fig. 5.10 Energy consumption—month distribution

Energy consumption versus env conditions. Environmental conditions are mainly associated with heating/cooling operation and thus the correlation with energy consumption should be depicted in the following, we show the accuracy of an energy modelling framework that consider environmental conditions (indoor and outdoor) in the analysis and following a long period training process (Fig. 5.11).

Energy consumption versus Device Type. It is evident that the consumption of the different building devices directly impacts total consumption (as total consumption is the aggregation of device level information). The disaggregation of consumption to the different device types is key information to better understand consumption patterns. Thus, correlation of device level consumption with total consumption is required to extract device level consumption patterns.

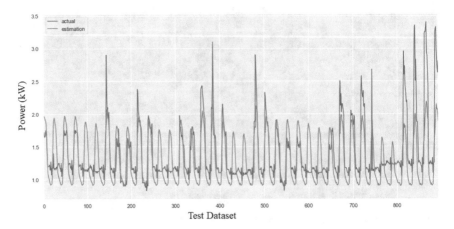

Fig. 5.11 Energy consumption versus indoor conditions

We presented in brief the different components and processes that consist of the Occupants Behavioural Profiling framework. This holistic framework is split into different modules to reflect the different layers of the analysis, namely occupancy, comfort, health and energy behaviour. For each of them, sub processes are defined by applying different separate methods and analytics techniques. This microservices based approach ensures a high level of modularity of the overall Occupants Behavioural Profiling framework. The engine is fully adaptable to the demo site needs and requirements and thus fine tuning of the model may apply to meet each project specific objectives and data availability.

Chapter 6
Digital Tools for Fast-Track Renovation Operations

Teemu Vesanen, Kiviniemi Markku, Kostas Tsatsakis, and Gabriele Masera

Abstract Digital tools for fast-track renovation operations developed in the project aim to shorten the duration of renovation and disturbance to the occupants with BIM-enabled methods and tools in operations management at site and with prefabrication to speed up the installation tasks. The chapter presents an ensemble of tools, concepts and use cases. First, two tools are described that are used to support construction production management and user communication. Then a concept how product data could be used as part of the tools and further how the product data and the tools could support in achieving the overall BIM4EEB objectives in the use cases of prefabricated exhaust air heat pump and prefabricated thermal insulation. Target of the work was to improve the state-of-the-art planning and monitoring. A new tool was created that combines the BIM model and typical work breakdown structure (WBS) based project scheduling into location breakdown structure (LBS) based user-interface. Continuously updated LBS provide valuable information to stakeholders with web-service and mobile applications. The 24/7 situational awareness of the renovation activities status provides unprecedented transparency of the project progress. Hence, the system allows scheduling the site activities with shorter lead times to shorten the total construction duration while it is possible to immediately take control of possible deviations in implementation. The reliable progress data is available also to the clients and occupants with right timed guidance and safety instructions. The other aim in the work was to utilise BIM for increasing the share of prefabrication in renovation projects. The BIM-based design allows to manage the compatibility and tolerances between design disciplines and adapting those with mapped geometry of the building will enable the prefabrication and preassembling of structural and

T. Vesanen · K. Markku (✉)
VTT Technical Research Centre of Finland Ltd., Espoo, Finland
e-mail: markku.kiviniemi@vtt.fi

K. Tsatsakis
SUITE5 Data Intelligence Solutions Limited, Limassol, Cyprus

G. Masera
Department ABC—Architecture, Built Environment and Construction Engineering, Politecnico di Milano, Milano, Italy

© The Author(s) 2022
B. Daniotti et al. (eds.), *Innovative Tools and Methods Using BIM for an Efficient Renovation in Buildings*, SpringerBriefs in Applied Sciences and Technology, https://doi.org/10.1007/978-3-031-04670-4_6

75

system components also in renovation. Two best practice examples were developed and described showing how to utilise prefabrication in real renovation scenarios.

Keywords Fast-track renovation · Prefabrication

6.1 BIMPlanner: A Tool for BIM-Enabled Construction Production Management to Plan and Track Site Operations

BIMPlanner is one part of BIM4EEB toolkit, a toolset for residential renovation projects that share data through a central BIM Management System (BIMMS). BIMPlanner is a cloud-based software for detailed planning and management of construction operations at site. The aim of the software is to share situational awareness of the construction activities at site to all participants of a renovation project. This information provides transparency of all site operations and improves productivity with better information enabling right timing of resource usage. The software allows to plan the site operations with more efficient throughput times and control the implementation with shared tracking of all activities.

The use scenario of the BIMPlanner is based on Location Based Management System (LBMS) that is a part of Lean Construction practices. Location-based management is a planning and control method that aims to optimize the schedule of project and at same time allow the continuous workflow (Kenley and Seppänen 2010). The site activities are planned and tracked by work locations creating workflow through those work locations. This will provide improved control of the workflows enabling faster proactive actions if delays are found.

In BIMPlanner the LBMS is applied with defining the work locations as reference to BIM objects which will define the workflows in relation to building coordinate system. For the indoor activities the work locations are defined with set of IfcSpaces. The work locations outside of the building are defined indirectly with reference other object types. Better approach would be to define these with IfcSpatialZones but such entities are not typically modelled in IFC, so in the current version of BIMPlanner use this indirect method on work location definition for external activities.

The major use scenario in BIMPlanner is the weekly scheduling of the ongoing activities by work locations and inputting the status data of the activities to store and share up to date tracking data (see on the right in Fig. 6.1). Before this weekly planning some input is needed for detailing the master schedule if needed. The master schedule is a high-level schedule agreed by the client and the main contractor as a plan for total construction period. The master schedule is imported into BIMPlanner to set targets for detailed planning, and master schedule timing cannot be changed in BIMPlanner. In the midterm planning (centre in Fig. 6.1), the need to divide master activities in sub-activities may appear for managing the scheduling of different works phases of the activity. Another part of this activity detailing is to plan and define the

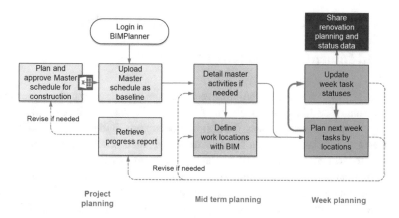

Fig. 6.1 Main functionalities in BIMPlanner

appropriate work locations for each activity as in weekly planning it is possible to plan each activity's timing separately for each work location.

One main goal in technical implementation of BIMPlanner was applying and testing some parts of the Digital Construction Ontologies (DiCon) developed in BIM4EEB.[1] In the BIMPlanner backend server the scheduling data and needed IFC data for work locations are stored as Linked Data in graph database according the DiCon ontologies. Since the project specific scheduling data and IFC data are confidential, their access is restricted to selected systems and users. The BIMPlanner is implemented as a server-client application where user interface is implemented in a browser and server component is communicating with other backend systems. Between BIMPlanner and BIM Management System (BIMMS) a linking layer called BIMLinker was implemented that maintains linking with BIMMS but also handles data storing/retrieving with graph database. The BIMLinker APIs were implemented as GraphQL[2]-based interface for higher level abstraction of the RDF-data. This GraphQL interface can be offered also to other software to access the scheduling data. For the research purposes the SPARQL endpoint is used also, and in current version BIMMS is retrieving the apartments related activity data with SPARQL query (access to SPARQL endpoint requires login). In this arrangement the BIMMS maintains the privacy issues with linking the apartments of the occupants to the corresponding IfcZones, and BIMPlanner handles only the IFC data.

---

[1] https://digitalconstruction.github.io/v/0.5/index.html.

[2] https://graphql.org/

6.2 A Building Occupants-Oriented User Interfaces Enabling Bi-Directional Communication and Information Exchange

One of the key objectives of the building management processes transformation is the incorporation of building occupants, inhabitants, and owners in the overall process. To ensure the active enrolment of the building occupants, it is mandatory to provide the necessary tools and services in order to get engaged. The main purpose of this section is to provide the details of a user-friendly web-based application dedicated to building occupants (i.e., Inhabitants and Owners), enabling them to be part of the building management ecosystem. At first, the building occupants should get informed in real time about the contextual conditions in building premises. Through the provision of context analytics dashboards, building occupants can get insights about their personal comfort preferences and further comfort related context analytics; as well as monitor their energy consumption on the way to raise occupant's awareness in energy efficiency and their buildings energy performance. Moreover, information about building processes (e.g., renovation actions, safety alerts) should be visible through the application.

On the other hand, and apart from the steer visualization of building related information, bidirectional communication flow should be supported to ensure the active enrolment of building occupants at building related decisions. More specifically, the building occupants should be able to provide input regarding their comfort preferences towards the extraction of more accurate comfort profiles, enhancing that way the personalization of the different services. In addition, user feedback about the different processes (e.g., renovation process negotiation, trigger safety alerts) should be supported by such tool. More details about the different functionalities are provided.

6.2.1 Real Time Building Management Information Visualization

This is the first view of the application focusing on real time information visualization about building contextual conditions. More specifically, the delivered application consists of three core features of added value for the building occupants', namely.

Visualization of both real time and historical information of individual comfort related parameters (such as indoor/outdoor, temperature, humidity, illuminance, and air quality metrics) supplemented with context analytics. Therefore, the users have access on actual environmental conditions but also on KPIs as extracted from analysis over historical data. Raw sensorial data are received from the BIMMS platform and send to the application where these are presented an intuitive manner. These raw sensorial data are further mapped to the static building related parameters to enhance end user visualization. Along with near real time information, end users of the tool can drill into the history and extract insights about environmental conditions profile. Moreover, as the provided application is intended for both owners and inhabitants dedicated views are provided serving its individual information needs, while also respecting their privacy. In more detail inhabitants can access both near real-time and historical information of its apartment's ambient conditions, while owners are able to see aggregate ambient condition information at building level. Indicative screenshots from the application are presented in Figs. 6.2 and 6.3.

Input provision from building occupants regarding their thermal and visual comfort status against set values of temperature **and illuminance in their apartment.** In general, building's occupants' feedback, is considered as a rich source of information towards evaluating environmental design practices and building operations. Therefore, an interactive user-friendly approach is provided to gather inhabitants' input through a five-point emoji-based scale. Inhabitants can select the appropriate emoji corresponding to their comfort status. A view of the approach considered for data gathering is presented in Fig. 6.4.

Correlation of inhabitants' feedback with ambient conditions data as gathered from the WSN is further considered towards extracting users' visual and thermal

Fig. 6.2 Real time environmental conditions

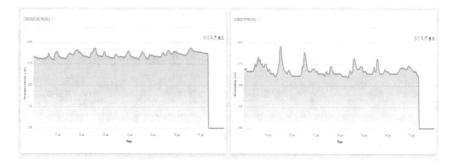

Fig. 6.3 Historical timeseries—environmental conditions

Fig. 6.4 User settings—environmental conditions

comfort preferences. An indicative view of the thermal and visual comfort boundaries of the users as extracted within the context of the project is presented in Fig. 6.5.

Visualisation of energy consumption (near real-time and historical) through a dynamic, user-friendly dashboard, supplemented with context analytics aiming at raising occupant's awareness on their energy consumption and buildings performance. This analysis is performed at the level of detail available from each demo site. More specifically, total consumption information, information from the different energy sources, information at device level may be available for visualization. In addition to timeseries information visualization, intuitive statistics are available for visualisation: Typical consumption profile (daily), Total period consumption/per source, etc. as presented in Fig. 6.6.

Fig. 6.5 User comfort profiles and boundaries

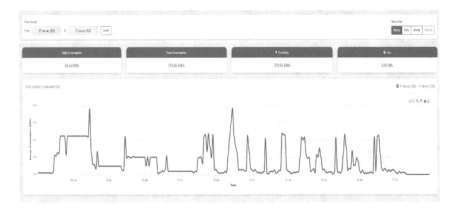

Fig. 6.6 Energy consumption profile

6.2.2   Building Processes Management Information Visualization

Apart from contextual information visualization, the building occupants should be also engaged on the typical processes performed at the building environment, namely: building renovation processes, building components updates, safety, and security events etc. Towards this direction, this 2nd view of the application should enable building occupants to periodically interact and co-design the different processes running at the building environment. Among the different processes, some key activities have been incorporated in the version of the application, namely:

Renovation Process Management and Negotiation: The building occupants should be always and conscious about the management and control of the renovation interventions. The application provides a service to control the renovation interventions, enabling them to actively participate (jointly with contractor) in the management of the intended renovation interventions. A view of the feature that provides renovation management activities overview is presented in Fig. 6.7.

Safety and Security Alerts and Notifications: Avoiding accidents on-site and preserving the H&S of the building's occupants is critical to any building management project. The application keeps owners and inhabitants of a building informed of the renovation activities by sending security and safety recommendations and/or Health and Safety (H&S) instructions regarding the ongoing and programmed renovation works. In addition, the end users of the app can report safety related situational conditions in the building environment during and post renovation process. A view of the feature that provides safety and security alerts and notifications overview is presented in Fig. 6.8.

Fig. 6.7   Renovation process management and negotiation view

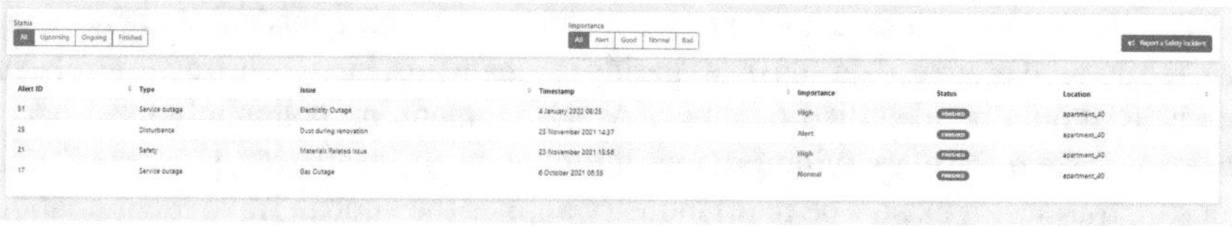

Fig. 6.8   Safety and security alerts and notifications view

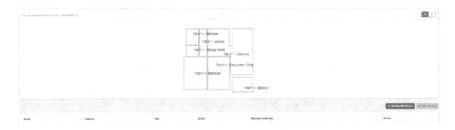

Fig. 6.9 Building components management and information enhancement view

Building Components Management and Information Enhancement: This service is about reporting of various updates on the building structures/elements within a building. The end users will be able to upload information in a structured way that is required to proceed with the fine-grained modelling of the building. On the other hand, information about the updates on building elements will be accessible by the different end users of the app, so they can get access to information that is valuable for the better management of the building structure. A view of the BIM components management and information enhancement is presented in Fig. 6.9.

Overall, the application for the building occupants is the tool to provide 24/7 situational awareness of the status of the building and its activities. The aim is to allow building occupants to become active stakeholders on the management of the building environment by providing tools and services that get the informed and further enrolled on building management related processes.

6.3 Managing Construction Product Data Using Linked Data Technologies

Currently, construction product data, which is handled during the design and construction phases of the building, are often stored as static documents (such as PDFs) in project-specific document repositories. The challenge with documents is that the data included in them are not semantically meaningful and require human browsing and interpretation. Thus, these static documents do not efficiently support the supply chain management of construction projects or the operation and maintenance phases of a property.

The starting point for managing product data in a machine-readable way necessitates structured product data and data sharing mechanisms. First, construction-related products should be identified uniquely. At the moment, no commonly used international system for classifying product information or identifying products in the construction sector exist. The retail industry applies the Global Trade Identification

Number (GTIN) that is provided by the international not-for-profit GS1 organisation. The construction sector should start utilising GTIN in identifying construction products more widely. Second, the product manufacturers should begin to use standardized product data templates to describe products.

The connection of product data to BIM models would enable advanced applications, such as digital twins, BIM-based facility management systems, building-related IoT in general, and even building management systems connected to BIM. The linking of product data to BIM also supports machine-to-machine workflows requiring machine-understandable product data. However, the recording of product data cannot be based, in general, to simple copying of all available data to a BIM system or other local repository, even though caching of such data will be useful if availability is critical. Product data itself can have many dynamic aspects, such as the emergence of replacement products, corrections of mistakes, recalls, links to vendors or pricing information, and usage data. Because of the dynamism, it would be useful to maintain the links between the BIM model and the source of product data, ultimately even to both directions.

The current Linked Data technologies already provide a working solution to the problem of connecting product data to entities specified in BIM models. Linked Data means a set of open technologies for decentralized data management, defined by the Web Consortium, including URIs, HTTP(S), RDF(S), and often also OWL, SPARQL, and Turtle. If the specification of a product and a corresponding BIM entity are both identified by a URI, it is a simple matter to represent a link between them in RDF, which is a standard model for data interchange on the Web. The link can be recorded by one party to provide a one-way access, both parties separately, or by a service providing access to both directions. Although such arrangements or commercial offerings do not exist yet, they are easy to envision from a technical perspective, apart from the complications created by access rights and systems security.

6.4   A Case of BIM-Enabled Exhaust Air Heat Pump Installation in the Project

An exhaust air heat pump has been selected as an example system as it has become a common solution for improving energy efficiency in renovation pro-jects. The limited example will encourage to implement BIM technologies even at a partial level in renovation projects and support in this way the market uptake targets and to create an example of how the BIM technologies will support the prefabrication of HVAC-installations and speed-up the operation time at the site.

Previous chapters present the construction production management and responsive user interface tools and a concept how product data could be used as part of those tools. This chapter and the next one about prefabricated thermal insulation element present use cases for those tools and product data.

The process for installing the exhaust air heat pump with the supporting BIM4EEB tools in the different phases of the project is shown below.

6.4.1 BIM-Based Order-Delivery Process of an Energy Renovation Project Using the BIM4EEB Tools

The installation of an exhaust air heat pump is expected to be profitable to building owner. A contractor drives the business case by identifying potential buildings/clients and offering the solution actively with cost–benefit based tendering. Currently one-of-kind construction project implementation may contain too much project-specific work and variation, concluding that such rather small projects may not be profitable for the contractor. To avoid this, the contractor needs to have pre-designed and configurable sets for main components of the system and a standard process for efficient implementation of the project.

Sales and marketing process includes activities for HVAC contractor to identify potential clients and marketing energy efficiency renovation services proactively. The HVAC contractor tries to identify potential sales opportunities for both existing and new clients with basic information on the buildings of the targeted clients. After the identification, the HVAC contractor contacts the client, and if the client approves to proceed with the initial contact, then proceeds to the next phase.

The **tender planning and contract phase** starts when the HVAC contractor inquires the client and collects the needed initial information for preparing the tender.

The AR fast mapping tool is applied to create a partial inventory model of needed spaces and HVAC installations. The contractor benefits from scanning the technical spaces, where the heat pump unit and heat storage tanks will be located, as well as routes for transfer of goods.

With the IFC-file from the AR fast mapping tool, the HVAC contractor studies the partial inventory model for building-specific boundary conditions and special designers design the product, i.e., the exhaust air heat pump. Energy experts generate renovation scenarios and estimate costs and benefits (OPR), applying BIMeaser.

The results gained by using BIMeaser are discussed with the client to make sure the client understands the impact of each renovation scenario in terms of owners' project requirements (OPRs). When the renovation scenario has been selected, a system-level design for tender is created. The system-level plan incorporates all major parts of the product that aims to reduce the energy consumption of the building. During the detailed design and prefabrication process phase, the system is detailed and documented for procurement, prefabrication, and installation at site.

Prefabrication has been found to increase the productivity of construction projects. In addition, previous research shows that prefabrication reduces the loss of building materials, improves the quality of construction and the safety of workers. Prefabrication can also be used to achieve cost benefits that result from good planning at different stages of the design and construction process, as lead times are shortened, design quality is improved, the site remains tidy, and logistics chains and installation efficiency will improve. (e.g., Bekdik et al.2016; Hanna et al. 2017; Khanzode et al. 2008; Wuni and Shen 2019).

After the contractor has received the detail design from the design consultant, the contractor defines and procures the system parts for the product package based on the design. Then, for the **prefabrication** the contractor sends orders to the manufacturer/pre-fabricator for the assembling of the product packages. Finally, the HVAC contractor receives the information that the product packages are is assembled.

At the beginning of the **logistics and installations phase**, the HVAC contractor plan the construction schedule and prepare operations at the site by procuring needed installation resources and planning site logistics. The contractor manages site operations with weekly planning and tracking site operations using BIMPlanner tool for documentation. The planning approach is following the basis of the location-based management method defined in Lean Construction literature (Tzortzopoulos et al. 2020). The plans and work status are shared to the client and occupant with BIM4Occupants tool. Work locations are defined with the BIM.

BIMPlanner tool sends schedule, and other relevant information to BIMMS which then sends the information to BIM4Occupants tool, which provides a guide for the building occupants during the on-site renovation works. The tool provides the occupants with safety alerts and insights into the ongoing and planned renovation processes. This enhances safety and security on-site.

During the **commissioning phase**, the renovation project and the installations are documented and commissioned. The HVAC contractor gives training to the client's facility management and maintenance personnel and/or service providers to use the system properly.

In case the HVAC contractor is the main contractor of the project, they are responsible for testing the systems and keeping a record of measurements. Inspections are arranged to ensure the system is installed and working according to regulations and deliver related documentation to the HVAC contractor. The HVAC contractor documents installations into the Digital Building Logbook of BIMMS.

The facility **operation and maintenance phase** focuses on monitoring the energy performance of the installed product and indoor conditions and making adjustments when needed. Client feedback data can also be collected and stored into BIMMS.

If the maintenance contract includes a promise for energy saving, a baseline for the energy has been set with some assumptions, and it is also important to monitor that those assumptions are still valid in the monitoring phase. For example, the energy-saving promise might assume that the indoor temperature in apartments varies between 21 and 25 degrees, but if some apartments are empty, having only 16 degrees for long periods of time, it could change the heat gain from the heat pump significantly.

6.5 A Use Case of BIM-Enabled Design for Prefabricated Thermal Insulation Components

This section presents an example of how BIM technologies can support a more widespread use of prefabricated thermal insulation components for the energy retrofit of existing buildings.

Considering the contribution this type of elements can offer to the EU's carbon neutrality targets, and their advantages over traditional solutions during the construction process, this chapter proposes a framework for the application of prefabricated insulation panels for energy retrofitting in a BIM environment. A broader application of BIM-based design tools, such as the guideline introduced in this report, can indeed support a more widespread adoption of prefabricated solutions among common envelope retrofitting options, allowing more informed decisions throughout the design and delivery process and a more structured engagement of the supply chain.

6.5.1 Modern Methods of Construction, Prefabrication, and BIM

The construction sector faces, among others, the double challenge of supporting the European Union's targets for energy efficiency and decarbonisation through solutions that allow to at least double the annual energy renovation rate by 2030 (European Commission 2020); and doing that while improving the efficiency of the process in terms of cost, time and use of resources (Hammond et al. 2014).

Modern Methods of Construction (MMC) are widely regarded as a promising solution to these challenges and, more in general, to many shortcomings of the construction industry. MMC are a broad term that encompasses different dimensions of innovation, such as off-site construction and related digital tools and techniques (MHCLG 2019). Prefabrication, often referred to also as off-site construction, is an approach to construction projects that seeks to move the construction process away from the site and take advantage of efficiencies from manufacturing approaches and standardisation.

The use of prefabricated panels for energy retrofitting—incorporating insulation, services, and finishing—can support the mentioned EU's goals about decarbonisation and increased renovation rates, while at the same time providing improved structural, thermal, acoustical, and architectural performances to existing buildings. Indeed, a wider use of prefabrication can be a game-changer for the construction industry, also for retrofit purposes, since it has the potential to reduce renovation times and ensure higher quality thanks to the industrialised fabrication of components (Sinclair et al. 2013).

A core aspect for the large-scale implementation of MMC, such as off-site façade panels for the energy retrofit of existing buildings, is their integration in a BIM

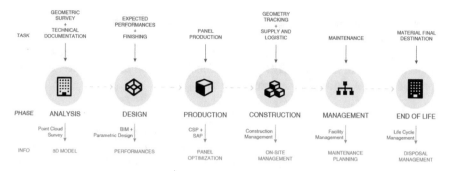

Fig. 6.10 Typical information flow for a panelisation process

approach and the consequent adoption of rich digital communication. Such an integrated approach requires that all data (finished models, geometric components and CNC—computer numerically control models, etc.) be effectively structured, properly managed and updated through all design and construction phases, therefore highlighting the importance of shared guidelines and standards to support the design and production process (BCA 2016; Alfieri et al. 2020). The research work on the state of the art, however, highlighted that the situation is currently rather fragmented, with several tools and guidelines mostly available only for specific aspects of the design and delivery process.

In the framework of the BIM4EEB project, therefore, a general-purpose guideline supporting decision-making in the early stage of design, when choices about construction technologies need to be made, was developed. Starting from the information gathered from an overview of existing prefabrication technologies, detailed experience from past research projects, and interviews with manufacturers, the whole life cycle of façade panels was articulated according to standardised phases with general applicability (Fig. 6.10).

6.5.2 Guideline on the BIM Implementation of Off-Site Façade Panels for Retrofit

Based on the abstraction of the typical information flow presented above, a process mapping of the design, fabrication and installation of façade panels was developed, leading to a guideline about the BIM implementation for the design of prefabricated thermal insulation components. The guideline highlights the correlations between the different project stages and identifies the actions necessary by the involved actors to optimise the workflow according to a MMC approach.

The whole process analysed refers to the design activities carried out by the design team (e.g., architect, cladding specialist, etc.) and is specifically targeted to the design and implementation of prefabricated insulation solutions on existing facades (i.e., a specific use case of MMC). Every effort was made to present an accurate description of the process, while remaining at a general level that is independent of the specific construction technology.

The guideline is articulated in nine stages, consistent with the structure of information adopted in the BIM4EEB project. Considering the limited space, only the contents of these nine stages are presented here.

Stage 1—Initiative

This stage defines the project scope and objectives formally. Targeted retrofitting actions on the existing building are defined, identifying the areas of intervention and verifying the regulatory feasibility of the intervention. Functional, aesthetic and cost requirements are collected from the Client. Finally, after a formal pre-contract BIM Execution Plan proposal, it is necessary to agree with the Client on the as-is modelling of the existing building to be renovated, and MMC adoption strategies. These guidelines need to be shared with all the Stakeholders. No model data is produced at this stage.

Stage 2—Initiation

Once the rules on data acquisition and BIM Execution Plan are formally defined, it is possible to proceed with a site survey and the subsequent modelling of the building; this can be done with laser scanning and/or other tools such as the BIM4EEB Fast Mapping Toolkit. This stage is crucial, producing a joint base for all future work. Technical, historical, and occupancy information are collected. After the definition of the MMC adoption strategies in Stage 1, an early assessment is conducted about the opportunities for their use. A detailed as-is BIM model of the existing building is developed.

Stage 3—Concept Design

An early-stage project is developed, verifying its coherence with the overall project scope and objectives. This stage includes geometrical and design explorations, using meta-technological objects for design optioneering purposes.[3] The Concept Design Stage is fundamental to understand the feasibility of the retrofit with off-site prefabricated insulation panels, identifying possible issues or limitations (both geometrical and energy-related).

Starting from the Stage 2 as-is BIM model and from several knowledge-based constraints from the designer (e.g., dimensions of the panels, possible integration with façade elements and installation, etc.), parametric placeholders' panels are added to the model, including only basic meta-technological information. The meta-technological panels are used for a first optioneering of the project. This optioneering phase is mostly based on geometric and architectural factors, while identifying possible issues for the panelisation and minimising the number of different panels. Information incorporated in the placeholder object for this stage are dimensions, area and volume and anchors. This step is useful to minimise unwanted design changes

---

[3] Optioneering is the commonly used term to describe the in-depth consideration of various alternatives and options to find the best or preferred alternative or option.

in subsequent stages and is obtained through the automated or semi-automated comparison between different proposals that can be easily shared with the Client for feedback.

Stage 4—Preliminary Design

The Preliminary Design Stage starts with the review of the Client's feedback on the different layouts previously explored in Stage 3. After a meta-technological option is confirmed, it is necessary to acquire performance targets, such as the U-value, to evaluate different technical solutions, each with its typical materials, are evaluated. Manufacturers can now be included in the process to provide the necessary information through early engagement; otherwise, available technical information or knowledge bases can be used. Each time the panels are modified, their technical data change accordingly, thanks to the parametric evaluation of the previous stages. This allows the team to export bills of quantity and using the model to show the stakeholder's feedback. Early datasheets containing dimensions, number of different elements, total number of anchors, and thermal performances are generated from the model for approval.

Stage 5—Developed Design

This stage aims to select a technical solution according to the Stage 4 Client's feedback. After the identification of a technical solution, several analyses and calculations are performed, starting a recursive process in collaboration with the Supply Chain. This includes performing energy, structural and LCA simulations. After the chosen solution is analysed and detailed, a first parametric time and cost definition can be shared for a budget validation phase. By the end of this Stage, after the budget is validated through early Contractor and Supply Chain engagement, detailed parts are studied and prototyped, and data-rich models generated. The Stage 5 BIM models can be shared with other disciplines for early coordination.

Stage 6—Detailed Design

In this stage, the digital model form Stage 5 is refined to incorporate inputs from the Supply Chain, producing a Construction BIM model. In this stage, time and cost are implemented into the Construction model, allowing the design team to produce a WBS.

An overall construction programme schedule and assembly sequencing are developed and can be shared with the Supply Chain to verify the construction process. This can be issued by developing fabrication and installation sequences, method statements and a resource management plan. This information can then be used in the BIM4EEB BIMPlanner tool to produce a lean and area-based representation of the construction works at each moment in time.

In this stage, there is a strong necessity to collaborate with the Supply Chain and manufacturers. In fact, most of the data needed for the development of the Construction model, such as the assembly sequencing, are delivered from the manufacturer itself. This exchange of information is the starting point for the production stage

(Stage 7.1, Pre-Construction), allowing a smooth transition between designer and producers.

Stage 7—Construction

The two phases of off-site fabrication and on-site assembly are strictly connected. For this reason, Stage 7 is subdivided into Stage 7.1—Pre-Construction and Stage 7.2—Construction. Façade panels are manufactured during the Pre-Construction stage. Time-related data embedded into the Construction model allows planning of just-in-time logistic to study delays and storage periods of panels.

Thanks to the cooperation with the manufacturers, shop drawings are integrated into the model. Transferring the manufacturer information into the model is crucial for the following stages.

During the On-site Construction stage, thanks to the BIM4EEB BIMPlanner and BIM4Occupants tools it is possible to track and share construction activities with the users in an all-in-one model considering the planned programme and assembly sequence, therefore significantly reducing disruption.

The output of the Construction stage is an As-built BIM model, fully integrated with all the information from the construction site and uploaded nearly in real-time as activities advance.

Stage 8—Building Use

During the Use Stage, the main objective is to update the as-built model with information about maintenance, facility management and, moreover, with the data coming from the BMS. The updated model is a Digital Twin of the physical building. With an accurate Digital Twin, it is possible to keep the ordinary maintenance plan up to date and to define the Life Cycle costs for the whole life cycle of the building. The model has to be integrated with the Facility Management system and the information about maintenance for each element.

Moreover, linking the BMS information into the model allows to identify possible issues and unwanted expenses in advance.

Stage 9—End of Life (EOL)

The end-of-life requirements are mainly related to the information from the manufacturer. Including this information into a regularly updated Digital Twin can help produce a fully developed EOL support documentation and, consequently, an EOL action plan.

Figure 6.11 shows graphically some of the data that is exchanged between actors of the process at the different stages of the process.

6.5.3   Test of the Guidelines on a Case Study

The guidelines summarised above were finally tested on a case study, to verify that the steps of the process and the information flow can actually be used in practice.

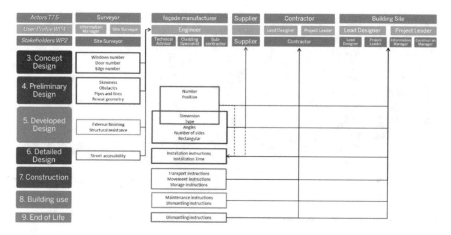

Fig. 6.11 Panelisation dataset and data exchange between various actors across the process stages

A simplified version of the Italian BIM4EEB case study project was used for this purpose.

The first part of the test focuses on the first stages of the guidelines (from 1 to 5), demonstrating the framework's effectiveness, and facilitating the selection of the configuration that best meets the requirements set by the Client. The test also aims at demonstrating the possibility, through inputs given by the data-rich BIM model provided in Stage 7 and by design constraints, to obtain coherent and integrated results in a design flow shared by the different project actors. The output (design of the prefabricated panels fitting the specific case study) contains the necessary information to interface with production and site operations. The second part of the exercise, interfacing with the BIMPlanner tool for the construction-related activities (Stages 6 to 9), demonstrates the possibility of its integration in the workflow and, at the same time, how the organisation of operations on site deriving from a prefabricated system is different from a traditional one. As part of the test, two alternative technologies for the prefabricated panels, namely TRM (textile-reinforced mortar) sandwich and timber frame, are considered. Only some highlights of the test will be presented in this section.

In Stage 3—Concept Design, the BIM model is acquired from the BIM4EEB BIM Management System, with some simplifications of the actual geometry for the purposes of this test. Thanks to parametric tools, a first round of optioneering is conducted about the geometry of the panels; in particular, the goal is to maximise the number of equal panels, minimise the total number of panels, and reduce or eliminate completely unique panels. Figure 6.12 shows the output of this process, which provides information such as total number of panels, the number of different panels to be produced, and a datasheet with the identification of individual elements.

In Stage 4—Preliminary Design, the performances of the solutions are evaluated thanks to the addition of technical information, allowing to move from meta-technological objects to a panel with more details. Once the preferred technological

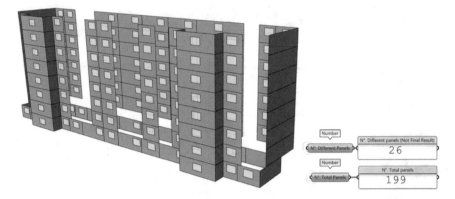

Fig. 6.12 First panelisation step with meta-technological panels, timber frame solution

solution is chosen, in Stage 5—Developed Design the IFC model including the panels is exported and shared with the manufacturer for validation. The panel layers are designed in detail, defining aspects such as the specific materials, their thickness and other properties, and anchors positioning. After further structural and energy analyses are completed, a fully verified and integrated BIM model is exported and ready to be shared with all the stakeholders, and in particular downstream with the manufacturer (Fig. 6.13).

Finally, in Stage 6—Detailed Design, the BIMPlanner tool is used to assess how the significantly shorter installation times allowed by the retrofit process with prefabricated panels impact the organisation of construction activities on site. The BIMPlanner tool requires the knowledge of the location and duration of each construction phase: in the case of prefabricated façade panels, it is therefore necessary to define their number for each front of the building and the related installation time. The number and location of panels are extracted from the BIM model obtained

Fig. 6.13 Perspective view of the BIM model with the selected panelisation

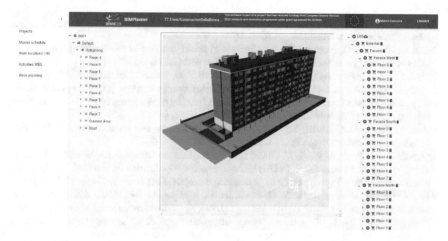

Fig. 6.14 BIMPlanner work location LBS page: upload of the fully panelised IFC model

in Stage 5, in the form of a spreadsheet also presenting their dimensions. Installation times are instead derived from a generalisation of available information from previous case studies and literature. The installation time for each panel is estimated based on its size, the number of anchors required (larger panels need more anchors), the length of perimeter joints, the connection details and its position in the façade (panels higher up in the façade imply longer lifting times).

After defining the necessary panelisation times for each façade, the compiled master schedule can be uploaded in the BIMPlanner tool; the BIM model is also loaded, with the façade panels as independent elements (Fig. 6.14).

After the WBS is associated on the Work Location LBS page of the BIMPlanner tool with specific elements, such as the individual panels, a fully articulated weekly planning is available.

At the end of Stage 6, thanks to the interoperability of the process with the BIMPlanner toolkit, a fully articulated weekly planning is available; a comparison of the panelised solution with a traditional ETICS insulation system shows a reduction of the installation time by 2/3 thanks to prefabrication.

References

Alfieri E et al (2020) A BIM-based approach for DfMA in building construction: framework and first results on an Italian case study. Architectural Eng Des Manage 16(4):247–269. https://doi.org/10.1080/17452007.2020.1726725

BCA (2016) BIM for DfMA (design for manufacturing and assembly) essential guide. Singapore building and construction authority and Bryden wood. Available at: https://www.corenet.gov.sg/media/2032999/bim_essential_guide_dfma.pdf

Bekdik B, Hall D, Aslesen S (2016) Off-Site prefabrication: what does it require from the trade contractor? Int Group Lean Constr 43:43–52

European Commission (2020) A renovation wave for Europe—greening our buildings, creating jobs, improving lives. Bruxelles, pp 1689–1699

Hanna AS, Mikhail G, Iskandar KA (2017) State of prefab practice in the electrical construction industry: qualitative assessment. J Constr Eng Manage 143(2):04016097–1/8

Hammond R, Nawari NO, Walters B (2014) BIM in sustainable design: strategies for retrofitting/renovation. In: Computing in civil and building engineering—proceedings of the 2014 international conference on computing in civil and building engineering, pp 1969–1977. https://doi.org/10.1061/9780784413616.244

Kenley R, Seppänen O (2010) Location-based management system for construction: planning, scheduling and control. Spon Press, London and New York

Khanzode A, Fischer MA, Reed DA (2008) Benefits and lessons learned of implementing building virtual design and construction (VDC) technologies for coordination of mechanical, electrical, and plumbing (MEP) systems on a large healthcare project. J Inf Technol Constr 13:324–342

MHCLG (2019) Modern methods of construction, introducing the MMC definition Framework. Available at: http://www.cast-consultancy.com/wp-content/uploads/2019/03/MMC-I-Padbase_GOVUK-FINAL_SECURE.pdf%0A, https://www.gov.uk/government/publications/modern-methods-of-construction-working-group-developing-a-definition-framework

Sinclair D et al. (2016) RIBA plan of work 2013 designing for manufacture and assembly overlay. Royal Institute of British Architects. Available at: http://consig.org/wp-content/uploads/2018/10/RIBAPlanofWorkDfMAOverlaypdf.pdf

Tzortzopoulos P, Kagioglou M, Koskela L (2020) Lean construction, core concepts and new frontiers. Routledge. ISBN 9780367196554, 460 p

Wuni IY, Shen GQP (2019) Holistic review and conceptual framework for the drivers of off-site construction: a total interpretive structural modelling approach. Buildings 9(5)

Chapter 7
Demonstration in Relevant Environments

Andrea Giovanni Mainini, Martina Signorini, Jaroslaw Drozdziel, Aleksander Bartoszewski, Sonia Lupica Spagnolo, Teemu Vesanen, Davide Madeddu, Eva-Lotta Kurkinen, Kostas Tsatsakis, Eoin O'Leidhin, and Markku Kiviniemi

Abstract Three building case studies were chosen with the purpose of demonstrating the BIM4EEB BIM-based toolkit. The selected buildings are both social houses and residential apartments respecting the needs of vulnerable inhabitants. To increase the representativeness of the test case the buildings are located in three different locations with different climatic conditions, specifically Italy, Poland, Finland. For all the case studies analysed, BIM models were created with different levels of detail (LOD), which, thanks to the interaction with the BIMMS, make it possible to create a common environment for the representation and use of the data collected and subsequently shared between the different tools. Among the three demonstration sites, the Italian site is undergoing building envelope renovation interventions such as the realization of the thermal insulation with ETICS technologies and the replacement of external windows. In order to test the different tools, a demonstration procedure has been defined for them, constituted mainly by workshop activities and quantitative and qualitative evaluations. To assess the level of accomplishment with respect to stated objectives and project success a validation methodology based

A. G. Mainini (✉) · M. Signorini · S. Lupica Spagnolo
Department ABC—Architecture, Built Environment and Construction Engineering, Politecnico di Milano, Milano, Italy
e-mail: andreagiovanni.mainini@polimi.it

J. Drozdziel · A. Bartoszewski
Prochem SA, Warszawa, Poland

T. Vesanen · M. Kiviniemi
VTT Technical Research Centre of Finland Ltd., Espoo, Finland

D. Madeddu
One Team S.r.l., Milano, Italy

E.-L. Kurkinen
RISE Research Institutes of Sweden AB, Gothenburg, Sweden

K. Tsatsakis
SUITE5 Data Intelligence Solutions Limited, Limassol, Cyprus

E. O'Leidhin
University College Cork (UCC), Cork, Ireland

© The Author(s) 2022
B. Daniotti et al. (eds.), *Innovative Tools and Methods Using BIM for an Efficient Renovation in Buildings*, SpringerBriefs in Applied Sciences and Technology,
https://doi.org/10.1007/978-3-031-04670-4_7

on Key Performance Indicators (KPIs) was delineated. Precisely, two categories of KPIs have been identified: "mandatory" and "secondary" addressing project objectives and in connection with the literature review and project use cases and tools. To calculate the KPIs standard baselines were estimated, such as are currently in an ongoing process to assess the traditional process that can be compared with the actual value associated with the BIM-based process. The chapter will present the methods and the first intermediate results of a demonstration process that is currently not yet completed and will later see a further application of the tools in dedicated demo sites. Environmental monitoring sensors were installed in selected apartments in Polish and Italian demo site, while were installed in common spaces for the Finnish building. Specific sensors set up have been analysed and chosen to fulfil the different needs related to the specific project outcomes. Inhabitants' availability, technical condition and flat exposition were criteria followed for the choice of apartments. Sensors allowed to improve the occupancy monitoring and to have a historical record of environmental values such as temperature, humidity and light strictly connected to users' preferences. The mobile application about renovation activities performed and residents' indoor home conditions—BIM4Occupants—has been installed by the users and specific workshops with inhabitants were carried out for registration purposes. The BIM Management System is currently collecting sensors' data stream and data stream between tools such as BIM4Occupants and BIMPlanner. Project monitoring and better communication among users were tested in a different workshop by applying the BIMPlanner tool in the plans and progress site operations. The functionalities of the refurbishment scenario simulation tool—BIMeaser—were tested in qualitative and quantitative design workshops respectively with the construction professionals using the two pilot sites in Italy and in Finland and with the aim of assessing the achieved time savings of using this tool compared to the manual data input process of the scenario simulation.

Keywords Demonstration · Best practices · KPI

7.1 The Adopted Methodology for Demonstrating the Application of the BIM4EEB Toolkit in Real Environments

To demonstrate the effectiveness of the developed BIM-based tools and the suggested methodology as well as the conformity to project objectives, BIM4EEB proposes three pilot sites located in different European countries. Italy, Poland, and Finland were chosen because of their different climate conditions to prove the applicability of the toolkit. Considering a user-centric approach and particular attention to vulnerable inhabitants, the selected buildings are both social houses and residential apartments.

The first step of the demonstration consisted in the sensors set up within the chosen apartments in order to retrieve environmental information coming from apartments use and address specific project needs. Subsequently, several workshops succeeded one another to test qualitatively and quantitatively the toolkit. Finally, a KPIs based methodology was defined to validate the success of project objectives. The demonstration activity is composed of qualitative and quantitative workshops that were and currently are carried out in the three pilot sites. To test the toolkit a generic demonstration strategy, followed by a KPIs demonstration procedure, has been defined. Building owners provided data to the tool developers, who run a first preliminary test, analysing how the tool works. At this point, a demonstration workshop occurred with a double function: from one side to explain the use of the tool to building owners and users, from the other side to do a secondary evaluation and verify and solve any complications. Once the tool demonstration activity is complete, KPI demonstration was performed. For this purpose, a study to define the baseline to calculate the KPI and to verify the success of the project has been developed.

Some examples of demonstration activities follow:

- data coming from sensors are visible in the BIM Management System and from other tools such as BIM4Occupants and BIMPlanner by the inhabitants and building owners.
- BIMPlanner has been presented to the partners since March 2021. By applying it in the plans and progress site operations is possible to improve project monitoring and communication. Several regular workshops have been organised with the building owner in the Italian demonstration site, to present the tool functionalities and to discuss possible implementations.
- BIM4Occupants has been successfully integrated with BIMMS platform to retrieve data from the Italian demonstration site. The demonstration of the tool to the inhabitants and building owners continues through workshops and support from technical partners.
- BIMeaser was tested in qualitative and quantitative design workshops with the construction professionals using the Italian and Finnish demo sites to assess the time savings by using this tool with respect to the manual data input process of the scenario simulation.
- Fast Mapping tool was tested in the three demo site in apartments and technical spaces.

As declared above, to test and verify the objectives of the project using the different applications, IoT devices and sensor systems have been selected and studied. Furthermore, several communication protocols such as Wi-Fi, Z-wave, Zigbee have been considered and chosen to meet the need. A hardware topology was defined to address the task requirements about:

- Indoor environmental conditions (temperature and RH, illuminance, occupancy/motion),
- Indoor Air Quality (IAQ) (CO_2),
- Electric power consumption.

Hence, different sensor devices were installed in the building case studies. A sensor hub with a connector and management function has also been installed to enable data integration from different sensors. This hub supports various communication protocols (Bluetooth, Z-wave and Zigbee), enabling hardware to data exchange and transmission. In addition, to collect the acquired sensor data, a system gateway was introduced. Sensors are linked to the gateway and due to the connection to its Application Programming Interface (API) data can be transmitted and stored into the BIM Management System, developed within the project, and performing the function of a data repository.

7.2 Description of the Demo Sites

The three Best Practice examples are residential buildings and social houses in Italy, Poland, and Finland. Description about the building itself and its main characteristics as well as the selection of apartments are presented for each specific pilot site. Renovation interventions, if present, are also outlined.

7.2.1 The Italian Demo Site

The Italian demonstration site is located in Monza, one of the biggest cities in northern Italy, in Via della Birona 47. The building, completed in 1981, is a residential social housing managed by ALER Varese—Como—Monza Brianza—Busto Arsizio (ALER VCMB), and it is structured on nine floors for sixty-five apartments in total. It is composed of plastered external facades and fair-faced concrete stairwells and fronts (Fig. 7.1).

The conservation status of the building is relatively poor. There are signs of loss of plaster and flooring as well as damage to the concrete cortical surface. To improve the performance of the residential building, two activities are being carried out. On the one hand, the realisation of insulating facade coat and the replacement of external windows with PVC windows are the main renovation interventions currently taking place. On the other, environmental sensor devices were installed to make apartments smarter and improve their performance.

The selection of individual apartments relied on residents' availability considering more cooperative families during the installation phase and data collection. Secondly, the choice was based on diversification related to exposure and different levels of the building (low, medium, high/ stairs A and B) and detailed analyses of the technical condition of individual apartments. Therefore, thirteen apartments have been individualised according to tenants' availability, exposure, and technical conditions. The selected apartments and the main characteristics of interest for the positioning of sensors are summarised in Table 7.1.

Selected apartments where to install sensors are two and three-room flats. The multipurpose sensors—measuring temperature, humidity, occupancy, and illuminance—were installed in the living room and the master bedroom, both in a two

Fig. 7.1 Italian demonstration site

Table 7.1 Main characteristics of each apartment

Apartment	Floor	Staircase	Apartment exposure	Type (2: two-room flat; 3: three-room flat)
A	Raised ground	B	East-west	3
B	Raised ground	A	East-west	2
C	Raised ground	A	East-west	2
D	3rd	B	East-west	3
E	3rd	B	East-west	2
F	3rd	A	North-east-west	3
G	6th	B	East-west	2
H	6th	A	East-west	3
I	7th	A	East-west	3
L	5th	A	East-west	3
M	7th	A	East-west	2
N	1st	B	East-west	3
O	6th	B	East-west	3

Table 7.2 Sensors set up in Italian demo site

Devices		Apartments					
		Common space	Two-room flat		Three-room flat		
		Single room space	living room	Bedroom	living room	bedroom	bedroom
Multisensor	Temperature	X	X	X	x	x	–
	Humidity	X	X	X	x	x	–
	Occupancy	X	X	X	x	x	–
	Luminance	X	X	X	x	x	–
IAQ Sensor	CO$_2$	–	X	–	x	–	–
	PM2.5	–	–	–	–	–	–
	PM10	–	–	–	–	–	–
	VOC	–	–	–	–	–	–
Radiator heat cost allocator		X	X	X	x	x	X
Thermostatic valve		X	X	X	x	x	X
Electric power meter		–	X		X		
GAS meter		–	–		–		
HUB		X	X		X		
Internet connection		X	X		X		

and three-room flat. The Indoor Air Quality (IAQ) sensor was placed only in the living room because it is considered the room of the house where the inhabitants spend most hours, assuring a more significant data collection. In the energy meter room, the electric energy meter was installed. Finally, one sensor hub for apartment was installed for acquiring data from different sensors. The electric power meters were then installed directly in energy meter rooms of the apartment and connected to dedicated hubs. Hence, sensors and meters were installed in the case study according to the distribution in Table 7.2 and Fig. 7.2.

7.2.2 The Polish Demosite

The polish pilot site is in the south of Poland in Chorzow town. The object of the study is a residential building built in 1902 characterised by 5 floors, 12 apartments and 3 commercial areas located on the ground floor, for a total of 1330 m^2. The data available come from different sources, both digital and on paper and from on-site inspections. Geometric and topological data about the generic aspect of the building for preliminary assessments are available, while data regarding apartments are extracted from original 2D drawings. The building itself would assess mainly

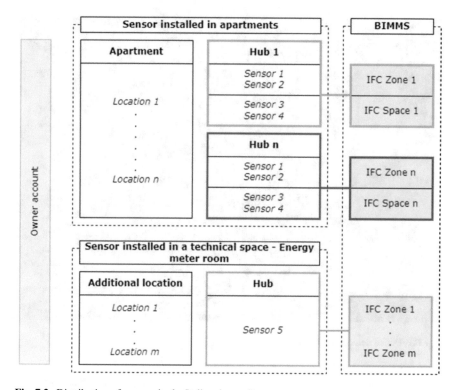

Fig. 7.2 Distribution of sensors in the Italian demo site

the design and planning functions of the platform, fast mapping capabilities and application of digital tools for HVAC design, operation, and efficiency management with the input of HMI and occupant´s profiling mechanism. Unlike the Italian demonstration site, no renovation involving construction work was planned or required (Fig. 7.3).

The selection of apartments for demonstration activities on the Polish pilot site was based on detailed analyses of the technical condition of individual apartments and according to the strong need for inhabitants to undertake further renovations (in some apartments, renovation works were carried out previously). Moreover, the availability of inhabitants and diversification of flat expositions were considered.

After site inspections, some general conclusions about the building technical state have been collected:

- The electric installation has cables that cannot stand the current, resulting in low possibilities of power supply. One apartment cannot have electric devices installed. The electric installation requires complete renovation in another one.
- Every shaft would need additional revision and cleaning, so that the natural ventilation would be possible. Three apartments have problems with shafts, and reverse thrust.

Fig. 7.3 Polish demonstration site

- Most of the apartments would require a change of boilers and distribution of domestic hot water.
- Radiators should be changed if working with lower temperature fluids.
- Piping, in general, is in good state, however, six apartments require a complete change of the system.
- There are two apartments that are unconditioned.
- Humidity issue in the apartments. The right wing of the building is suffering major humidity problems, with mould in four apartments.

In polish case study mostly two bedrooms apartments were selected for the installation of the sensors. Equipment installations consider inhabitants preferences, basic presence, and technical limitations of devices. The Indoor Air Quality (IAQ) sensors were installed in the living rooms. Multipurpose equipments measuring temperature, humidity, occupancy, and illuminance were installed in the bedrooms. Electric energy meters were installed inside main technical shafts which combine meters dedicated to each apartment. Sensors, hubs, and meters were installed in the case study according to the Table 7.3.

Table 7.3 Sensors set up in Polish demo site

Devices		apart A	apart B	apart C		apart D		apart E	
		living room	living room	living room	bed-room	living room	bed-room	living room	bed-room
Multi-sensor	Temperature	X	X	x	–	x	x	x	X
	Humidity	X	X	x	–	x	x	x	X
	Occupancy	–	X	–	–	–	x	–	X
	Luminance	–	X	–	–	–	x	–	X
IAQ sensor	CO_2	X	–	x	–	x	–	x	–
	PM2.5	–	–	–	–	–	–	x	–
	PM10	–	–	–	–	–	–	x	–
	VOC	–	–	–	–	–	–	x	–
Electric power meter		–	–	x		x		X	
GAS meter		–	–	–		–		–	
Heat cost allocator		X	X	x		x		X	
HUB		X	X	x		x		X	
Internet connection		X	X	x		x		X	

Fig. 7.4 Finnish demonstration site

7.2.3 The Finnish Demo Site

The Finnish pilot is located in Espoo, in Southern Finland near Helsinki. The actual building. which was originally built in 1992, is a five-floor apartment building owned by Finland's largest pension provider KEVA.[1] Keva invests its funds also in real estate with the responsibility focus on property energy efficiency. Keva's goal is to become carbon neutral by the end of 2030 in terms of carbon emissions from the energy used in its direct real estate investments.

The building covers a total of 4100 m^2 floor area. The building was brought under the BIM4EEB project as an energy renovation project from Caverion. The demonstration pilot for the BIM4EEB project was changed in the middle of the project due to COVID-19 pandemic caused a delay in the first apartment building renovation project that did not fit into BIM4EEB schedule. Some of the tools of the BIM4EEB project were tested on the earlier pilot building. BIMeaser was used in energy simulation for selection between different heating system upgrade options that could have been installed on the first building (Fig. 7.4).

After the pilot building change, the heating system of the new building was upgraded with the addition of an exhaust heat recovery system. The goal of the renovation project was to extract the excess heat from the HVAC system to enhance the performance of the district heating system during the heating season. The planning and installation works were done in collaboration with Caverion and a subcontractor Tom Allen Senera Oy. The renovation works and testing of the system were conducted during Q3-Q4 of 2021. All the installation works were done on the facility areas with no work in apartments due to privacy and health reasons during the pandemic. The progress of renovation works was monitored with weekly online meetings between Caverion, VTT and Tom Allen Senera Oy. Different phases of the installation works were collected and reported to BIMPlanner tool to assess the

---

[1] https://www.keva.fi/en/this-is-keva/rental-flats-and-business-premises/.

Fig. 7.5 Finnish demo site boiler room with the new heat exchanger

planned and actual progress of the project. A demonstration day to test the Fast Mapping tool with RISE and Caverion was held to model an IFC file of the boiler room where the installations were made. The project on site was finished in January after final inspections (Fig. 7.5).

7.3 BIM4EEB KPIs Framework and Validation Methodology

The Key Performance Indicators (KPIs) are used to estimate the level of success of the project and assess the achievement of planned and defined objectives. The selection considered KPIs that follow the needs of the project and can assess its results.

KPIs are divided in "mandatory" and "secondary". Mandatory KPIs are based on BIM4EEB objectives, and they are in line with the stakeholders' requirements. They need to be addressed during the last phase of the project. While the Secondary KPIs relate to the literature analysis and the review of project use cases and tools.

Moreover, considering the different interests of the project from the renovation process to the energy performance passing through users' comfort, BIM4EEB KPIs have been grouped in the following six categories: "renovation process", "energy performance", "comfort", "economic", "social", "environmental".

"Renovation process" category is related to the specifics of the renovation process such as time and cost, "energy" is connected to the energy requirements and performance of the renovated buildings such as energy consumption, energy saving, life cycle assessment parameters, "comfort" refers to occupants thermal, visual and building acoustics comfort, "economic" KPIs regards the feasibility of buildings renovation regarding cost, life cycle costs, project profitability and LCC assessment, "social" KPIs investigate stakeholders' such as designers, construction companies—FMs, occupants) perception and "environmental" category is related to CO_2, CO, PM, VOCs.

Finally, not all KPIs are applied to the developed BIM-based tools. They are addressed according to the peculiarities of the single tool. Consequently, since not all developed BIM-based tools have been tested in the three demo sites of the project, the KPIs are different according to the considered demo site too (Fig. 7.6).

For KPI calculation, a precise formula has been defined in the project deliverables and associated with each KPI in order to be measured. Hence, it is necessary to calculate specific parameters to assess the level of success. Especially, a baseline,

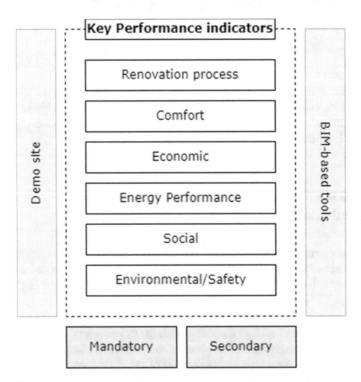

Fig. 7.6 KPIs framework: KPIs grouped in mandatory and secondary, and according to demo site and BIM-based tool

which is a reference parameter corresponding to the traditional renovation process, has to be taken into consideration in the most relevant cases.

For the sake of clarity an example of mandatory KPI is proposed. Among "renovation process KPIs", Renovation Time Reduction is a KPI defined to assess the time saving occurred during the renovation process due to improved management of the renovation activities.

7.4 Tools Application in Real Environments

At the date of writing this manuscript, the process of demonstrating the tools is still in progress, which is why only a few indicative results are given, outlining the main processes identified and tested, as well as some of the results obtained.

Below is a breakdown of these contents among the tools that are part of the project.

7.4.1 BIMMS

The BIMMS functionalities were tested and demonstrated in the Italian, Polish and Finnish demonstration sites. The BIMMS was used to store documentation, models, and drawings along with the continuous data streaming coming from IoT devices installed in the dwellings (Fig. 7.7).

The BIMMS' SPARQL Endpoint was available as an interface to share and query the project data to the authorised stakeholders and helped to improve the knowledge

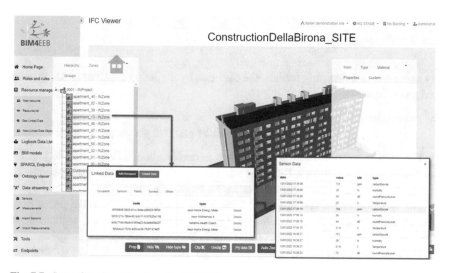

Fig. 7.7 Some functionalities from the BIMMS

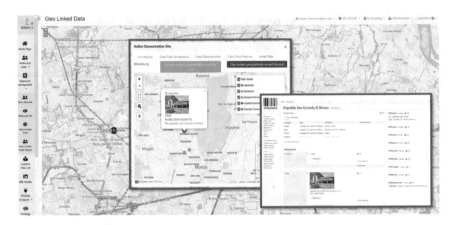

Fig. 7.8 Geo linked data of the Italian demo site

about the surroundings (environmental, social, and administrative data) retrieving additional data used to enrich the BIM models (Fig. 7.8).

The demonstration activities allowed to use and test the REST API interfaces to update interactively the IFC BIM model demonstrating that the technology is ready and can support faster workflow that can avoid some bottleneck tasks (export-upload-download-import) typical of traditional file exchange methods.

The stakeholders evaluated the BIMMS and gave positive feedback stating that the user interface has an intuitive design and the functionalities provided allow them to exchange information, monitor the construction works, and manage the building data in more effective way compared to a traditional renovation approach.

7.4.2 Fast-Mapping

By using the fast-mapping toolkit, the user can by the help of a laptop, AR glasses and a sensor stick:

1. map what's inside the construction,
2. create IFC-files out of the mapping results and real geometry in 3D.

The mapping result might be used when to decide about a renovation or just for documentation. The created IFC-files will be used in the BIMMS system and/or as an illustration when discussing renovations with stakeholders.

The function of the fast-mapping toolkit was tested in the different demo sites (Italy, Finland, and Poland). In Italy one empty and one occupied apartment were used and in Finland an equipment room and in Poland an occupied apartment.

The demonstration included:

1. scanning the considered rooms by an Imager 5010.

Fig. 7.9 From left, 1. laser scanning, 2. Point cloud downloaded to the laptop, the program used is named companion app, 3. Mapping the wall in the room by the sensor stick

2. creation of point clouds out of the scanned rooms which were downloaded to the laptop and then transferred to the AR glasses. The AR glasses was a HoloLens2.
3. opening the point cloud in the HoloLens2 and align it with the reality.
4. mapping the rooms with the sensor stick.
5. create IFC-files in the HoloLens2 out of the reality in combination with the point cloud (Fig. 7.9).

7.4.3 BIMeaser

The BIMeaser test results have shown that the modelling is faster with the digital BIM assisted process compared to the manual modelling. The digital and traditional approaches were compared in the testing. The easy application of renovation measures and reviewing the results contribute positively to the overall modelling time in BIMeaser process. The overall increase in modelling time reduction is more than 75% with BIM assisted process compared to manual modelling. In addition, the BIM-assisted BIMeaser process also brings data accuracy and helps to avoid modelling errors, which are quite common in the manual approach (Table 7.4).

7.4.4 BIM4Occupants

As stated above, the building occupants should be able to get insights about contextual conditions in building premises, personal comfort preferences and further comfort and energy-related analytics. In addition, the building occupants oriented framework

Table 7.4 Time savings compared to traditional energy simulation in two real world case studies

Process phase	Modelling time for each phase Via Birona building, Italy (man-hours)			Modelling time for each phase Tapettikatu building, Finland (man-hours)		
	BIMeaser process	"Manual data input" -process	Reduction %	BIMeaser process	"Manual data input" -process	Reduction %
Starting point	3	3	0%	1	1	0%
Input data check and preparation for the energy scenarios	0.25	8	−97%	0.5	2	−88%
Use case 1. The build-up of the "As-is" energy and indoor climate model	3.6	19	−81%	2	10	−80%
Use case 2. Applying the renovation scenarios	0.5	1	−50%	0.5	3.5	−86%
Use case 3. Reviewing the impact of scenarios in terms of Owners Project Requirements	0.1	0.5	−80%	0.1	0.5	−80%
Total (man-hours)	7.5	32	−76%	3.9	17	−77%

allow building occupants to receive notification and alerts on ongoing works, receive safety hints and information (e.g. to avoid specific areas where works have not been finished yet), while on the other hand, enabling them to upload information that might be requested ad-hoc by contractors or any other relevant input they may consider useful, thus contributing to the constant and collaborative updating of BIM and as-built documentation. Indicative screenshots from the demonstration activities of the project are presented in the following figures (Figs. 7.10 and 7.11).

The outcome of this tool will further facilitate the building stakeholders to get insights about occupancy and comfort conditions in premises as well as properly plan the renovation activities of the project. Indicative results from the analysis performed at demo premises is presented in the following figures (Figs. 7.12 and 7.13).

The following are some of the Key insights:

Fig. 7.10 Renovation scheduling view

Fig. 7.11 Example of safety alerts

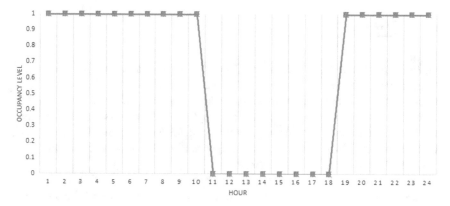

Fig. 7.12 Occupancy profiling results

1. Post processing and data handling of sensor data is required.
2. The analysis can extract useful results in terms of occupancy and comfort that
   may lead to:

   a. Better simulation of energy and building performance.
   b. Better operation of the building environment.
   c. Increase user comfort and productivity.

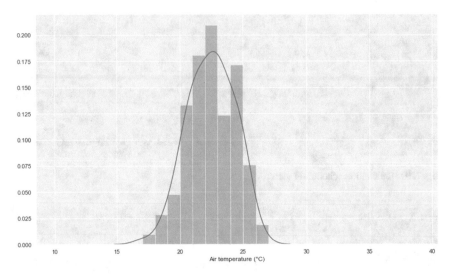

Fig. 7.13 Thermal comfort profiling results

3.   While the framework was tested at residential premises, the same applies also for office buildings and thus the overall framework is replicable.

7.4.5 BIMcpd

BIMcpd stands for BIM Constraint Checking, Performance Analysis and Data Management. The tool is a user-friendly self-intuitive software suite designed to reduce processing time for constraint checking, increase building energy performance knowledge, standardise building energy data and help users to make informed and better decisions. The software is developed for Building Services Designers, MEP engineers, M&V Practitioners, Energy Auditor, Building Energy Managers, ESCO's, Research Performing Organisations.

BIMcpd contain three distinct intuitive applications that will allow users to (a) find recommended positions for HVAC, lighting, and other devices; (b) manage the data that they have on the above and create new data sets that they can share with other tools and (c) analyse data from sensors, energy bills, and other sources.

The Constraint Checking Tool is designed to extract information from an IFC file and compare it with data stored in the BIMcpd database, which allows the values from the model to be checked to ensure they are compliant with existing building codes, such as ventilation or lighting regulations (Fig. 7.14).

The BIMcpd Data Management tool allows for the upload of data from import files in which the fields can be mapped in order to standardise the way that the data

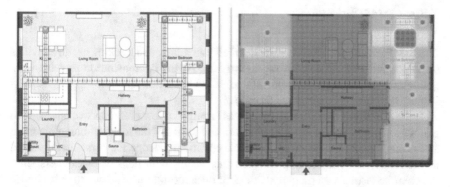

Fig. 7.14 HVAC Layout in BIMcpd Constraint Checking Tool

is being imported. Alternatively, data can be imported automatically from precon-figured sensors, and this was the method used in importing data from the demo sites.

Finally, the Performance Evaluation Tool follows the Measurement and Verifica-tion (M&V) process and principles, allowing data uploaded using the Data Manage-ment tool to be modelled. It allows for data manipulation, such as data filtering and outlier detection to increase the accuracy of the results (Fig. 7.15).

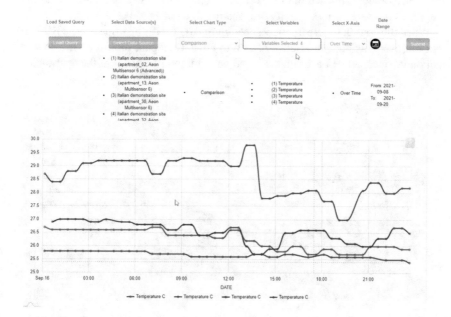

Fig. 7.15 Comparison of temperature across 4 apartments from Italian demo site

7.4.6 BIMPlanner

The BIMPlanner tool was tested for detailed scheduling and tracking of site activities in Italian and Finnish demonstration projects. As the demonstration projects were quite different in renovation content and extent the BIMPlanner testing focused on slightly different topics in these projects.

In **Italian demonstration project** was arranged a few common workshops with major project partners to introduce the tool and basics of the location-based planning and management method. With this background knowledge the personnel involved in the renovation project implemented the testing with help of researchers from Polimi. The testing was partially hampered by some external factors, mainly Covid-19 lockdown, and some material shortages in Italian market during renovation. The testing mainly concentrated on two activity types: façade renovation and internal windows exchange.

In the following is presented an example of window exchange planning including communication with the occupant of an apartment. In BIMPlanner the activities can be scheduled by work locations which means that e.g., the window exchange activity can be given planned times independently for each apartment. In preparation of the actual scheduling, the work locations are defined in a specific BIMPlanner view called Location-breakdown-structure (LBS). Figure 7.16 shows an example of the definition of work location "apartment H" (alloggio H) and in 3Dview are shown the IFC space objects attached to this work location. Those IFC-objects are the linking identifiers to share correct activity information to other BIM4EEB tools.

When the contractor creates new scheduling data for the apartments, the BIMMS tool retrieves this data and transfers data to BIM4Occupant tool to be shared with individual occupants of apartments based on common identifiers of the apartments.

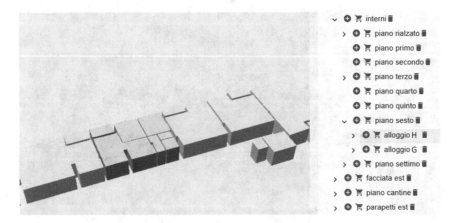

Fig. 7.16 Example of work location definition of "apartment H" with linked IfcSpace-objects in 3D-view

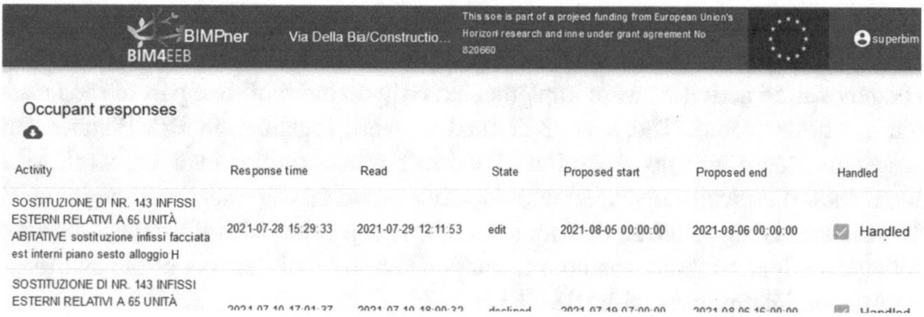

Fig. 7.17 Example of occupant's proposal of new timing for the activity in specific "apartment H"

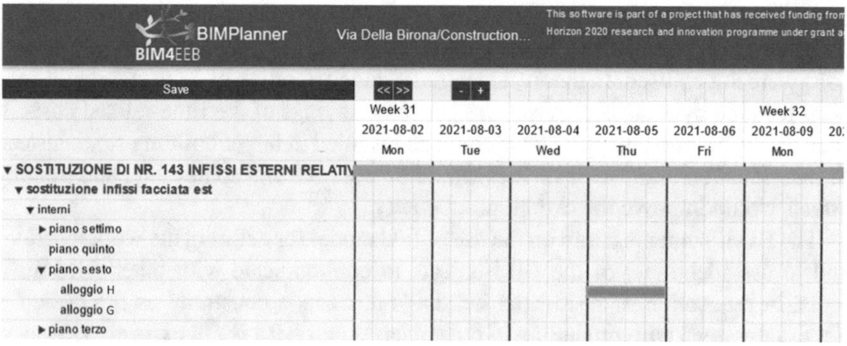

Fig. 7.18 Example of the final schedule for "window exchange" in "apartment H"

The occupant may decline the contractor's initial schedule and propose new timing for the activity. In Fig. 7.17 is an example where the occupant has proposed new timing for window exchange within two days' timeslots.

After the contractor receives the occupants' response the rescheduling will be done and new timing shared with the occupant. In Fig. 7.18 is an example of final scheduling for one certain day for window exchange in this apartment.

The testing in Italian demonstration project indicated some needs for the further developments of BIMPlanner and possibilities to test some other planning scenarios. The definition of the work locations was manual work although it needs to be done only once. If the apartments are defined systematically in IFC-data and can be extracted from other spaces and zones of the building those can be converted more automatically as work locations in the location-breakdown-structure of BIMPlanner. At minimum this would require guidance on what apartment identifiers shall be used in modelling and developing automated interpretation functionality in BIMPlanner. In the planning sense an alternative approach could be tested: the contractor could propose the initial schedule with a longer timeslot at floor level (i.e., same time period for all apartments in the floor) and let occupants to select the specific times for their apartments within a given timeframe.

The **Finnish demonstration project** was a specific energy renovation with the installation of an exhaust heat air pump in a residential multistorey building. The renovation activities were implemented only on the roof, one part of the façade and the boiler room. Those were defined as work locations in BIMPlanner. But in general, there was no repetition of the different activities over the work locations that are typically managed with location-based management method in which BIMPlanner is based on. In this demonstration project the BIM4Occupant tool was not tested either so there was no information sharing with tenants of the apartment or commercial premises with BIM4EEB tools.

However, BIMPlanner was used in project management and on weekly basis was created or refined next week's plan and recorded the actually started and finished activities. The testing revealed the fact that the total schedule was not tight and contained lead times that were reserved by the subcontractor to minimize the possible schedule risks. During implementation there were no major delays and the schedule risks were not realised. In theory, the initial planned duration of 13 weeks could have been reduced by around 3 weeks and get some savings of the time-related costs. If such energy improvement systems could be installed in larger building stock instead of individual building such savings would be possible and achievable, and training effects would improve the overall productivity.

The BIMPlanner is based on the use of IFC model for defining the work locations and to use identifiers of the BIM-objects to communicate with other BIM4EEB tools. In renovation of residential buildings this sets a constraint as the design is not usually done with modelling in typical cases and BIM is not existing. However, BIMPlanner does not require a detailed design model. In the Finnish demonstration project, the BIM was created with a reasonable amount of work by reading pdf-floor plans in modelling tool and raising those in 3D (Fig. 7.19). Some additional work was needed in placing e.g., windows and doors in place and adding some data like room and apartment identifiers in the model. Such a simplified model can be and was used also for BIM-supported energy simulations with BIMeaser tool.

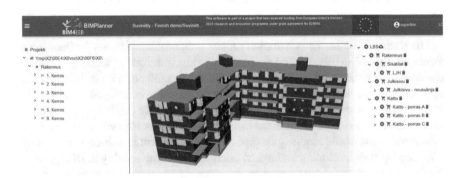

Fig. 7.19 Simplified BIM of the finnish demonstration project

7.5 Lessons Learnt

The focus of the project is on the digitization of methods and processes related to the construction sector, hence the proactive involvement of different stakeholders even if with minimal digital skills, but open to learning is required.

In this context, any individual involved in the process has to be included in dissemination activities and introduced to the use of tools and systems in order to stimulate personal or professional curiosity in tool use, highlight the advantages and increased possibilities of the new proposed approach, lay the foundations for a relationship of trust. A satisfied user is a key factor for a successful proposal and this success can be measured through the number of accesses to the tools as well as the quality of the results that can be produced. The BIM4EEB dealt with different stakeholders involved in the process, both professional (i.e., whose purpose in using the tool is related to the optimisation of the work process) and users who gain personal benefit from the application of the tool as the inhabitants of the demo sites.

The process of installing, managing, and integrating the tools must be straightforward from the very first start, encouraging user's autonomy in managing the information. A preliminary feedback from the questionnaires submitted to professional users shows general ease of use of the tools, even if technical support is not completely excluded. The increased data interoperability among the developed tools is also intended to reduce the time and cost of the activities. Hovever, it is true that some tools (e.g., scanning and modelling tools) require both equipment and specific skills that need dedicated training and users' competence improvable by continuous practice.

Regarding time and cost management, the application of lean construction practice processes requires weekly or bi-weekly planning that should be actively supported by the general contractor and the other stakeholders involved. An effective approach involves a tool able to support the designer, but at the same time able to guarantee flexibility in managing the organisation of activities and sub-activities. This is particularly true when the process requires the presence at the site of sub-contractors assigned to dedicated tasks. However, it is essential to formalise the information framework and the flow of data, but from the owner's point of view this is reasonable only if added value of the up-to-date digital data of the site operations can be identified. These data should reduce the need for control resources and costs and/or result in better decisions for project control. The compensation of providing these data to client is even more pressing in cases of uncertainty periods such as the one related to COVID-19 and shortages of construction materials particularly in the Italian case. The content and characteristics of a renovation project may vary considerably, and construction management methods shall be applied accordingly. In full scale residential renovation projects common methods that are used in construction in general can be applied. In limited energy improvement interventions when there is no full-time site manager and workers of different disciplines operate at site separately, these construction planning and control methods may not be the best approach. In such cases creating an application of remote workforce management methods used for example in facility management services should be considered.

A measure of the quality of the proposed interventions, as well as the state of the buildings involved in the project, can be known by the building owners thanks to the environmental sensors installed inside the flats.

To assess the environmental variables, it is necessary to have dedicated and appropriate types of sensors that guarantee hardware interoperability, and a have a correct position in order to obtain representative measurements of the context. For this purpose it is necessary to keep in mind how occupants' preferences for the installation, as well as the presence of fixed furniture, can influence the repeatability of the positioning within the different housing units.

Focusing on the inhabitants, although there has been a general good acceptance of the installed devices, the use of single multi-sensors devices able to measure different environmental variables is recommended, allowing a minimum perceived invasiveness inside the flats. Wherever possible, a battery installation is preferred by the inhabitants, both to avoid a burden on consumption and on the availability of sockets in the rooms. One of the requirements, particularly true under COVID-19 restrictions, was also to minimise the number of interventions and accesses inside the flat, but this kind of installation requires an additional maintenance burden to replace the batteries.

Concerning the population of occupants and adding the social housing context, it is clear that we are dealing with an elderly population with limited technological and digital knowledge, as well as devices suitable for the purpose. A longer introduction process is therefore necessary, defining periods of support and interaction and data management assistance.

The presence of real-time alerts and feedback also improves both the interaction and the user's understanding of the data, even if they require an additional effort from the building owner side. The possibility provided to the occupants to communicate their availability regarding the renovation activities that will take place in their flat, as well as the possibility to report any inconvenience due to the process, has been shown to improve the communication relationship between building owner and resident. However, this process requires attention from the building owner to ensure a prompt response and management of the file.

As the toolkit is based on BIM technology, efficient use of the tools could improve by broadening the knowledge in this field by renovation process participants.

Conclusions

BIM4EEB project has fostered renovation industry developing an attractive and powerful BIM-based toolkit able to support designers in the design and planning, construction companies to efficiently carry out the work, and service companies to provide attractive solutions for building retrofitting.

Additionally, public, and private owners have validated by using the toolkit, which eases decision making and asset management, thanks to the exploitation of augmented reality and the use of updated digital logbooks.

BIM4EEB has delivered an innovative common BIM management system with linked data and a set of tools.

BIM4EEB BIM based toolkit is a basic instrument for increasing semantic interoperability between software and stakeholders involved along the overall renovation process (design, planning, construction, operation, and maintenance.

End-users of the entire renovation process have actively participated to the development phases ensuring the full matching of project deliverables with the supply chain expectations and maximizing the value of what produced. In particular, one public administration and two general contractor companies have validated the toolkit in a social housing building in Italy and 2 private residential buildings in Poland and Finland, with relevant reductions of time, cost and energy consumption Inhabitants have taken advantage by the increase of building performances, quality, and comfort.

A focused market analysis and exploitation strategies have assured suitable market strategies.

The BIM4EEB tools are available standalone, as a whole or in different combination "as a service" usable in consultant services or by other Information Technology business area, if required, they are easily integrated in other existing Software environment thanks to predefined interfaces.

The developed best practices and guidelines for BIM adoption for public administrators and private stakeholders have been finalized to support the wide uptake of the project results and ensuring the acceptance of BIM4EEB toolkit by all stakeholders.

© The Editor(s) (if applicable) and The Author(s) 2022
B. Daniotti et al. (eds.), *Innovative Tools and Methods Using BIM for an Efficient Renovation in Buildings*, SpringerBriefs in Applied Sciences and Technology,
https://doi.org/10.1007/978-3-031-04670-4

Printed in the United States
by Baker & Taylor Publisher Services